Membrane-Assisted Crystallization Technology

Advances in Chemical and Process Engineering

ISSN: 2045-0036

Series Editor: Stephen M. Richardson *(Imperial College London, UK)*

Published

Advances in Chemical and Process Engineering – Vol. 2

Membrane-Assisted
Crystallization
Technology

Enrico Drioli
University of Calabria
Institute on Membrane Technology,
Italian National Research Council (CNR-ITM), Italy

Gianluca Di Profio
CNR-ITM, Italy

Efrem Curcio
University of Calabria, Italy

Imperial College Press

ICP

Published by

Imperial College Press
57 Shelton Street
Covent Garden
London WC2H 9HE

Distributed by

World Scientific Publishing Co. Pte. Ltd.
5 Toh Tuck Link, Singapore 596224
USA office: 27 Warren Street, Suite 401-402, Hackensack, NJ 07601
UK office: 57 Shelton Street, Covent Garden, London WC2H 9HE

Library of Congress Cataloging-in-Publication Data
Drioli, E.
 Membrane-assisted crystallization technology / Enrico Drioli (University of Calabria, Italy),
Gianluca Di Profio (CNR-ITM, Italy), Efrem Curcio (University of Calabria, Italy) World Scientific.
  pages cm. -- (Advances in chemical and process engineering ; v. 2)
  Includes bibliographical references.
  ISBN 978-1-78326-331-8
  1. Membrane reactors. 2. Crystallization. 3. Membranes (Technology). I. Di Profio, Gianluca.
II. Curcio, Efrem. III. Title.
  TP248.25.M45D753 2015
  660'.28424--dc23
                                                                    2015010205

British Library Cataloguing-in-Publication Data
A catalogue record for this book is available from the British Library.

In-house Editors: Thomas Stottor/Sree Meenakshi Sajani

Typeset by Stallion Press
Email: enquiries@stallionpress.com

Contents

Preface

Continuous sustainable industrial growth might be realized today and in the future with important innovations in process engineering.

The process intensification (PI) strategy well represents the contribution that process engineers can offer to sustainable industrial growth. The basic principles and objectives of PI are not simple and not easy to be realized nor to be accepted. The aim of PI is to bring about drastic improvements in manufacturing and processing, substantially decreasing equipment size and energy consumption, increasing plant efficiency, reducing capital costs, increasing safety, minimizing waste production, and reducing environmental impact. Sustainability and competitiveness are essential objectives for the process industry, and the accelerated implementation of PI will help to achieve a sustainable and economically strong process industry. The potential benefits of PI for the process industry are significant in terms of energy savings, reduction of CO_2 emissions, and enhanced cost competitiveness. They will significantly impact each sector of the process industry in one way or another.

Membrane engineering has already provided interesting solutions to some of the major problems of our modern industrialized society. Membrane processes meet the requirements of PI because they have the potential to replace conventional energy-intensive techniques, to accomplish the selective and efficient transport of specific components, and to improve the performance of reactive processes. Membrane techniques are essential to a wide range of applications including the production of potable water, energy generation, tissue repair, pharmaceutical production, food packaging, and the separations needed for the manufacture of chemicals, electronics, and a range of other products. Membrane technology has already gained a huge importance in the last two decades and now is competing with other separation technologies in terms of energy efficiency, high separation capacity, selective separation, and capital investments. On a number of occasions, commercial conventional separation processes in industry were converted to membrane separation processes. This is what happened, for example, in water desalination; in this field, conventional thermal technologies have been gradually replaced with reverse osmosis (RO). Membrane engineering has a much wider spectrum of potential applications as unit operations in process engineering than in other technological areas. Membrane operations can be used to conduct molecular separations (microfiltration, ultrafiltration, reverse osmosis, etc.), chemical transformations (membrane reactors, catalytic membranes, membrane bioreactors, etc.), and mass and energy transfer between different phases (membrane contactor, membrane distillation, membrane crystallizer, membrane emulsifiers, membrane strippers, membrane scrubbers, etc.)

Crystallization is one of the most widely applied separation processes in the chemical industry. It also play a central role in many scientific and technological sectors. Numerous products in daily use, such as additives for hygiene and personal care, pharmaceuticals, fine chemicals, pigments, and several other foodstuffs, are formulated as crystalline powders. This is because the solid state affords enhanced stability for storage and easier product management by users. In some technological fields, as for example, in microelectronics, nonlinear optics and sensoring, the solid state is the base in the realization of semiconductor single-crystal devices; furthermore, crystalline materials are widely used in heterogeneous catalysis in chemical industry, in the controlled release of active substances, as separation media in chromatography, and, more generally, in the nanotechnologies. In scientific research, single crystals or crystalline powders are required in structure-based investigations, with particular emphasis to medical advancement by rationale drug-design strategy and to structure-based materials' development.

Generally, macroscopic crystal properties have a remarkable impact with respect to their effective use. In order to obtain a high-quality product, it is crucial to control product properties such as crystal morphology (shape, habit, average size, size distribution), structure (polymorphism), and purity (regular arrangement of the building blocks into the lattice). All these properties have a considerable impact on the final use of a crystalline product. Preferred crystal shapes are required, e.g. for more efficient downstream operations such as recovery by filtration from mother solutions, drying, compression in producing tablets, and storage.

In heterogeneous catalysis, small, highly mono-dispersed in size, and uniformly shaped crystals are better suitable to achieve the highest surface-to-volume ratio and increased catalytic efficiency. In the pharmaceutical industry, the different polymorphic forms of the same substance have their characteristic physical, chemical and biological properties, making each of them a diverse and patentable drug.

In the case of proteins, crystals of adequate size and with elevated order in their crystalline lattice are needed for structure determination at the atomic level by X-ray diffraction analysis.

However, despite the importance of crystallization operation in many fields, current approaches still suffer from some limitations that affect both the products' quality and their process efficiency, especially when taking into account the more recent applications of crystallization process in industrial and technological fields. The main limitation is poor reproducibility in the final crystal characteristics, which is associated with limited supersaturation control, imperfect mixing, reduced and inhomogeneous distribution over the plant of solvent removal or anti-solvent addition points, and reduced possibility to modulate the supersaturation generation rate. In addition, the high energy requirements for heating in conventional evaporators or pumping in vacuum systems are important limitations in conventional crystallizers. Consequences of these limitations are the difficulties in polymorphism selection, enantioselective crystallization, and co-crystallization in pharmaceutics; the crystallization of bio(macro)molecules and other complex biomolecules in life sciences and biotechnology; the effective recovery of valuable products from waste streams and brines coming from heavy industrial lines.

If, on the one hand, continuous strengths are addressed to understand the basic mechanisms of crystallization aiming to increase control over the final properties of the materials produced, the recent developments in industrial crystallization stimulates more efforts to improve strategies and tools, and to develop new equipment by which crystallization could be operated in a well-controlled way.

On these bases, in the past few years, the interest in combining membrane operations and solution crystallization is motivated by the aim to develop more efficient crystallization processes. Accordingly, membrane technology might represent a valid and innovative tool to introduce significant improvements in crystallization. This approach has been put in practice in a number of forms of membrane-assisted crystallization (MAC) configuration.

In this book we will present the efforts — in large part still in progress — for introducing a new design for one of the most important unit operations: crystallization. Membrane crystallizers represent well the tentative implementation of a new innovative design of a basic important unit operation based on membrane engineering, which will satisfy most of the requirements of the PI strategy in terms of principles and objectives.

It is interesting to consider that this new design offers not only solutions to problems at the limits of the traditional ones, but also new opportunities, in part still unexplored, for facing new problems. The cases of specific polymorph crystallizations, the control of the kinetic of crystal growth, and the faster crystallization of biomolecules of high molecular weight are all typical examples of benefits of membrane crystallization processes: production of much higher-quality products, competitiveness with other existing technologies, and implementation on an industrial scale.

Chapter 1

Overview of Membrane-Assisted Crystallization Operations

1.1 Introduction

The interest in combining membrane operations and solution crystallization is motivated by the aim to develop more efficient crystallization processes. Membranes, with their intrinsic characteristics of efficiency and operational simplicity, high selectivity, and permeability for the transport of specific components, compatibility between different membrane operations in integrated systems, low energetic requirement, good stability under operating conditions and environmental compatibility, easy control and scale-up, and large operational flexibility, represent a technology that well satisfies the concept of the rationalization of chemical productions, leading to significant innovation in both processes and products. Accordingly, membrane technology provides significant improvements in crystallization. This approach has been put in practice in a number of membrane-assisted crystallization (MAC) configurations. The main features of MAC are:

1. the use of membranes as precision devices to control the composition of the crystallizing solution the level of supersaturation and its generation rate by tuning the mass transfer across the membrane;
2. the action of the porous surface of the membrane as suitable support to activate heterogeneous nucleation mechanisms;
3. the possibility to induce nucleation and crystal growth in separate compartments, thus reducing the risk of membrane clogging even when the membrane works as heteronucleant;
4. the ability to affect solid features such as size, shape, polymorphic form, and purity;

5. easy scale-up or scale-down processes;
6. reduced energy consumption with respect to cooling or evaporative crystallization.

Several membrane operations are used to realize MAC: reverse osmosis (RO), membrane distillation (MD), osmotic distillation (OD), forward osmosis (FO), nanofiltration (NF), microfiltration (MF), membrane contactors (MCs), membrane reactors (MRs), supported liquid membranes (SLMs), and membrane templates (MTs).

In this chapter, an overview of MAC techniques will be given. The description of the MD/OD-MAC, which is the main subject of this book, will be given in full detail in the next chapters.

1.2 Timeline of the Development of Membrane-Assisted Crystallization Processes

Dialysis can be considered as the first membrane operation used to perform crystallization,[1] prevalently used in the 1960s to crystallize biomacromolecules with high molecular weight. Since then, the occurrence of crystallization phenomena associated to other membrane operations has frequently been observed. However, to date crystallization occurring in membrane operations was considered as a side effect, because crystal formation on the membrane surface or inside the pores, dramatically reduced the flux and process productivity,[2–5] with premature degradation of the membrane itself.[2,6] Accordingly, initial interest in crystallization and membranes was principally motivated by the necessity to adopt strategies to avoid solid formation and precipitation on the membrane surface.[7]

A further application of a membrane unit as a crystallizer dates back to 1986, when RO hollow fibers were used to simulate the early stages of stone formation in the renal tubules due to precipitation of calcium oxalate.[8–10] In 1989, the possibility to crystallize a solute by using MD at sufficiently high concentration was roughly envisaged.[11] In 1991, this same concept was put into practice by using MD to recovery taurine from pharmaceutical wastewater streams in the form of crystalline material.[12] In the same year, FO was used to promote a controlled crystallization of biomacromolecules[13–19] in an osmotic dewatering system.[20] Also, MF was used to increase the concentration of a suspension of foreign seeds, up to the required level for seeded crystallization, for the softening of drinking water.[21] In 2001, the terms membrane crystallizer and membrane crystallization (MCr) were coined.[22] The new technology refers to the extension of the MD or OD working principles to a process where crystal nucleation and growth is carried out, starting from an undersaturated solution, by adjusting solution composition (supersaturation) by solvent removal in vapor phase through porous membranes. In more recent years, MAC operations based on NF[23] for the concentration of a stream prior to batch cooling crystallization have been studied. This method has been extended also to the crystallization of poorly water-soluble active pharmaceutical ingredients (APIs) using organic solvent nanofiltration (OSN) membranes.[24]

As a further development, a technique in which an antisolvent is pressed directly in liquid phase through a MF membrane towards a crystallizing solution (or vice versa) to produce solid particles with narrow crystal size distribution (CSD) has been proposed.[25–27]

In a different approach, the extension of the MCr concept to antisolvent operation, where a porous membrane is used to dose the amount of antisolvent in the crystallizing solution, has been explored.[28] Here, solvent composition within the crystallizing solution is controlled by diffusion in vapor phase, in accordance with the general concept of MCr and dissimilarly from the configuration in which the liquid phase directly permeates through the membrane.

MCs, MRs, and SLMs have also been used to implement MAC processes. $BaSO_4$ particles were obtained by reaction crystallization after a solution of a $BaCl_2$ had been pressed through a membrane into a K_2SO_4 solution.[29] The resulting supersaturation induced nucleation and particles grew on the lumen side of hollow fiber membranes.[30,31] SLMs containing a mobile carrier were used for the synthesis of $CaCO_3$ crystals exhibiting different structures.[32]

To summarize, MAC operations can be classified on the basis of their different working principles as:

(1) MD/OD-based processes, where solvent molecules evaporate from the feed side, diffuse through a porous membrane, and recondensate on the opposite side, generating supersaturation in the crystallizing solution, under the action of a gradient of vapor pressure as driving force;

(2) Hybrid processes, in which pressure-driven membrane operations (RO, NF, and MF) are used to concentrate a solution by solvent removal in liquid phase, while the solute is retained by the membrane, and crystals are recovered in a separate tank, often operated at lower temperature and with seeding. This also includes crystallization operated in organic solvents by OSN membranes;

(3) MAC, where an antisolvent is forced — directly in the liquid state — into the crystallizing solution (or vice versa) through the pores of a membrane under a pressure gradient and, as long as it diffuses in the crystallizing solution,

Table 1.1.   Timeline overview of MAC processes.

Year	Membrane Operation	Reference
1968	Dialysis	1
1986	Reverse Osmosis	8,9,10
1989	Membrane Distillation	11,12
1991	Forward Osmosis	20
1996	Microfiltration	21
2001	Membrane Crystallization	22
2002	Membrane Reactors	29
2004	Supported Liquid Membranes	32
2004	MF antisolvent	25,26
2005	Nanofiltration	23
2009	Antisolvent Membrane Crystallization	28
2014	Organic Solvent Nanofiltration	24

exchanges with the original solvent thus producing the abrupt abatement of solubility and solute precipitation;

(4) Antisolvent MCr, where dosing of the antisolvent in the vapor phase inside the crystallizing solution is carried out by means of a membrane, according with the same working principle described in Point 1;

(5) Membrane-assisted reactive crystallization, where two solutions containing different reactants or components are contacted, in liquid state, by means of a porous membrane, so that a solid crystalline product is separated by after reaction active or passive diffusion in the reaction medium through the membrane. This category includes MAC processes based on MCs, MRs, and SLMs.

Table 1.1 summarizes the timeline development of MAC processes.

1.3   Dialysis

Zeelen and Wierenga[33] described the use of dialysis membranes for the crystallization of biological molecules on the micro scale. In this technique, diffusion and equilibration of precipitant molecules through a semipermeable membrane is used to slowly approach the concentration at which the macromolecule crystallizes (Fig. 1.1). Provided that the precipitant has a low molecular weight (like for a salt or an alcohol), it can easily penetrate the dialysis membrane, while the protein is slowly brought into equilibrium with the precipitant solution so that it precipitates by salting-out effect. The transfer of solutes and solvent through semipermeable membranes provides the slowly varying conditions required for

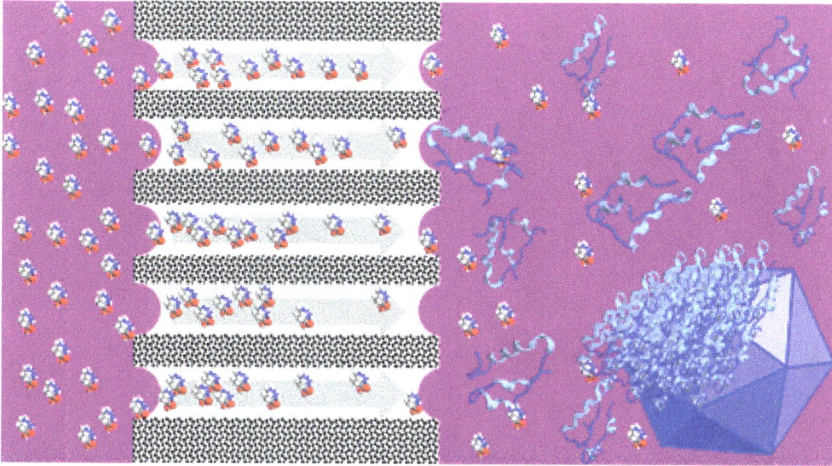

Figure 1.1. In dialysis, a low molecular weight precipitant easily penetrates a semi-permeable membrane and equilibrates into the protein solution, slowly approaching the concentration at which the macromolecule crystallizes by salting-out effect. The high molecular weight of the macromolecule prevents its migration to the opposite side of the membrane by size exclusion mechanism.

early crystal nucleation and for relatively undisturbed lattice formation of crystal growth.[1,34,35] Typically, membranes with molecular weight cut off (MWCO) around 10,000 are used in this configuration. They are not useful for the crystallization of molecules with lower molecular weight, because water and solutes travel through them simultaneously.

1.4   FO

Although FO has recently attracted interest for water pre-treatment, it was used for crystallization many years ago.[20] In this MAC method, originally called "osmotic dewatering", a non-porous membrane is used to separate two solutions having different solute concentrations:[36,37] a low concentration solution, containing the substance to be crystallized, and an extractant solution of salt at a high concentration (draw solution). Due to the osmotic pressure resulting from the difference in osmolarity across the membrane, only water diffuses down its chemical potential gradient from the low-concentration solution, permeating through the membrane into the high-concentration side (Fig. 1.2). Dewatering of the solution containing the solute increases the concentration and causes nucleation and crystal growth. The dehydration rate can be controlled on the basis of the flux of water across the

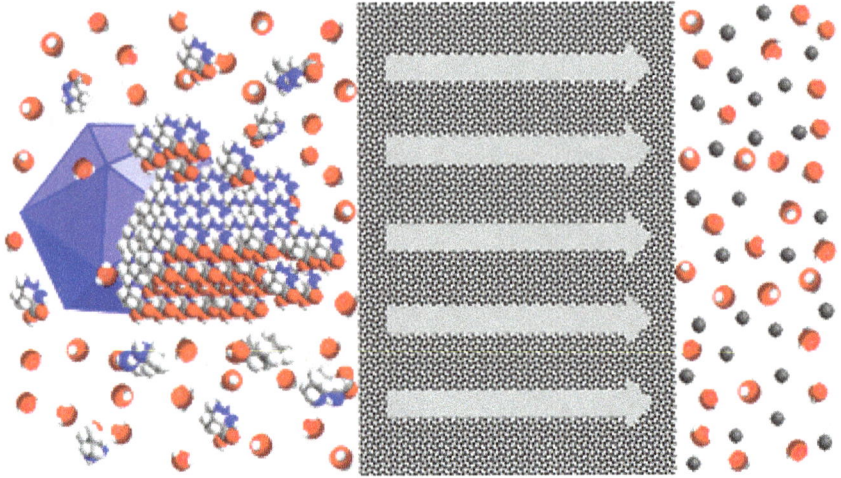

Figure 1.2. Concept of osmotic dewatering process: water is removed from the crystallizing solution in liquid phase through a non-porous membrane, under a driving force generated by the difference in osmolarity across the membrane.

membrane that is determined by the relative concentrations of all solutes inside and outside the membrane. Such adjustments, for example, to allow nucleation to occur very slowly,[38] while crystal growth can be augmented at later times by increasing the rate of solvent removal, thus avoiding the excess of nucleation (smaller crystals) and reducing the frequency of amorphous precipitation.

This method has been implemented in a multi-chamber device and a "fluid processing apparatus", a chemical transfer system designed for automated experiments in space.[36] It was successfully applied to grow triglycinesulphate, high-quality lysozyme crystals diffracting to 1.73 Å,[20] and an oligonucleotide[36,37] showing a maximum resolution of 2.46 Å, comparable to that of crystals grown by conventional methods.

1.4.1 *Related equations*[20]

In absence of externally applied forces, water flows down its osmotic gradient from the crystallizing solution to the draw solution. For ideal solutions, the total pressure drop across the membrane, ΔP, is given by the sum of the gravity head and the osmotic pressure expressed by the van't Hoff:

$$\Delta P = \rho g \Delta h + RT \Delta C, \tag{1.1}$$

where ρ is the density, g is the gravitational constant, Δh is the piezometric head, ΔC is the total osmolarity difference across the membrane, R is the gas constant, and T is the absolute temperature. Since a typical initial osmotic pressure is around 2 MPa, while the contribution of the difference in fluid head is around 20 Pa, the gravity term is generally neglected.

The flux of water driven across a membrane by osmotic pressure is governed by the Darcy's law:

$$\frac{1}{A}\frac{dV}{dt} = \frac{RT}{\mu R_m}\sum_{i=1}^{n}(\Delta C)_i, \tag{1.2}$$

where A is the cross-sectional area of the membrane, V is the volume of water that has permeated at time t, n is the number of solutes, μ is the viscosity of pure water, and R_m is the overall membrane resistance. Defining

$$(\Delta C)_i = C_{ir} - C_{iw}, \tag{1.3}$$

where C_{ir} is the total osmolarity of species i in the draw solution and C_{iw} is the total osmolarity of species i in the crystallization solution, at time t

$$C_{ir}(V_{r0} + V) = C_{ir0}V_{r0} \tag{1.4}$$

$$C_{iw}(V_0 - V) = C_{iw0}V_0, \tag{1.5}$$

where C_{ir0} is the initial total osmolarity of species i in the draw solution, C_{iw0} is the initial total osmolarity of species i in the crystallization solution, V_{r0} is the initial draw solution volume, and V_0 is the initial crystallizing solution volume. Rearranging:

$$C_{ir} = \frac{V_{r0}}{V_{r0} - V}C_{ir0}, \tag{1.6}$$

$$C_{iw} = \frac{V_0}{V_0 - V}C_{iw0}. \tag{1.7}$$

Combining Eqs. (1.3)–(1.7):

$$(\Delta C)_i = \frac{V_{r0}}{V_{r0} + V}C_{ir0} - \frac{V_0}{V_0 - V}C_{iw0}, \tag{1.8}$$

$$\sum_{i=1}^{n}(\Delta C)_i = \frac{V_{r0}}{V_{r0} + V}\sum_{i=1}^{n}C_{ir0} - \frac{V_0}{V_0 - V}\sum_{i=1}^{n}C_{iw0}. \tag{1.9}$$

Moreover:

$$\sum_{i=1}^{n}C_{ir0} = C_{r0}, \tag{1.10}$$

$$\sum_{i=1}^{n}C_{iw0} = C_{w0}, \tag{1.11}$$

where C_{r0} is the total osmolarity of all solutes in the draw solution and C_{w0} is the total osmolarity of all solutes in the crystallizing solution. Therefore

$$\sum_{i=1}^{n} (\Delta C)_i = \frac{V_{r0}}{V_{r0} + V} C_{r0} - \frac{V_0}{V_0 - V} C_{w0}. \tag{1.12}$$

Combining Eqs. (1.2) and (1.12), gives

$$\frac{dV}{dt} = \frac{ART}{\mu R_m} \left(\frac{V_{r0}}{V_{r0} + V} C_{r0} - \frac{V_0}{V_0 - V} C_{w0} \right). \tag{1.13}$$

Rearranging the above equation:

$$\frac{V_0 V_{r0} - V_{r0} V + V_0 V - V^2}{(C_{r0} - C_{w0}) V_0 V_{r0} - (C_{r0} V_{r0} - C_{w0} V_0) V} dV = \frac{ART}{\mu R_m} dt. \tag{1.14}$$

If α is the ratio of original crystallizing solution volume to original draw solution volume, $\alpha = V_0/V_{r0}$, dividing the above equation by V_0/V_{r0}, and by rearranging:

$$\frac{1 - (1 - \alpha) V_0 V - \alpha \left(\frac{V}{V_0} \right)^2}{(C_{r0} - C_{w0}) - (C_{r0} - C_{w0}) \left(\frac{V}{V_0} \right)} dV = \frac{ART}{\mu R_m} dt. \tag{1.15}$$

By separating the above equation into three parts and integrating them separately, with the initial condition that at $t = 0$ is $V = 0$:

$$\left[\frac{(C_{r0} - C_{w0})^2 \alpha V_0}{(C_{r0} - C_{w0}\alpha)^3} + \frac{(1 - \alpha)(C_{r0} - C_{w0}) V_0}{(C_{r0} - C_{w0}\alpha)^2} - \frac{V_0}{C_{r0} - C_{w0}\alpha} \right]$$

$$\times \ln \left[1 - \left(\frac{C_{r0} + C_{w0}\alpha}{C_{r0} - C_{w0}} \right) \left(\frac{V}{V_0} \right) \right]$$

$$+ \left[\frac{(C_{r0} - C_{w0})\alpha V_0}{(C_{r0} - C_{w0}\alpha)^2} + \frac{(1 - \alpha) V_0}{C_{r0} - C_{w0}\alpha} - \frac{V_0}{C_{r0} - C_{w0}\alpha} \right] \left(\frac{V}{V_0} \right)$$

$$+ \frac{\alpha V_0}{2(C_{r0} - C_{w0}\alpha)} \left(\frac{V}{V_0} \right)^2 = \frac{ART}{\mu R_m} t \tag{1.16}$$

All of the factors in the transport equation are constants or measurable values based on experimental conditions. Generally, the draw solution volume is much larger than the crystallization solution volume, $V_{r0} \gg V_0$ and $\alpha \to 0$, so the above

equation simplifies to

$$\frac{\mu R_m V_0}{ART C_{r0}} \left\{ \left(\frac{V}{V_0} \right) - \frac{C_{w0}}{C_{r0}} \ln \left[1 - \left(\frac{C_{r0}}{C_{r0} - C_{w0}} \right) \left(\frac{V}{V_0} \right) \right] \right\} = t. \qquad (1.17)$$

1.5 Pressure-Driven Membrane Processes

In this method, both the retentate (r) and the permeate (p) sides are in liquid phase. A pressure difference (ΔP) is applied over the membrane (porous or non-porous), resulting in a chemical potential difference ($\Delta \mu$), which is the driving force for mass transport. Only solvent (water) molecules are effectively transported through the highly selective membrane while solute molecules, that are rejected from the membrane by size exclusion or Donnan exclusion effects, accumulate on the retentate side (Fig. 1.3).

Among the different pressure-driven processes, RO was first used as a crystallization tool in hollow fibers configuration.[8–10] Experimental evidence for narrow CSD and occurrence of an unexpected pseudo-polymorphic form (dihydrate instead of monohydrate calcium oxalate) was obtained.[9] Keeping the pressure difference constant along the hollow fiber wall provides constant generation of the supersaturation along the tubes. As a result, a high degree of uniformity in the crystal habit and crystal morphology can be achieved.[8]

The following section gives a description of the most general theoretical approach used to describe a MAC process implemented by pressure-driven

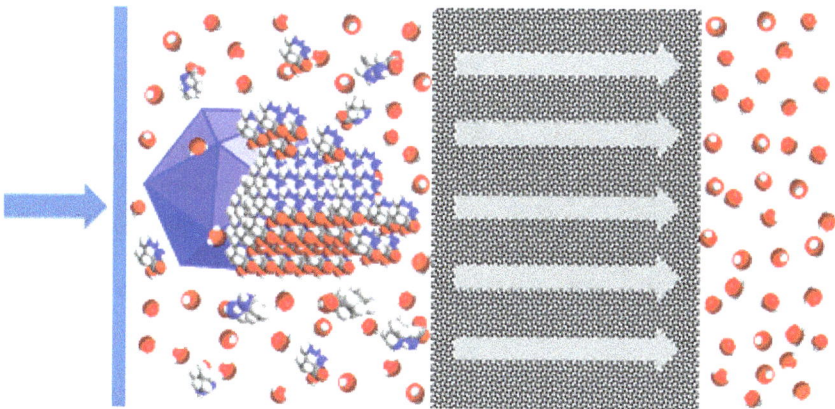

Figure 1.3. Working principle of the RO-assisted crystallization process a non-porous membrane is permeated by solvent molecules (water), while solute concentrates at the feed side, under a gradient of applied pressure.

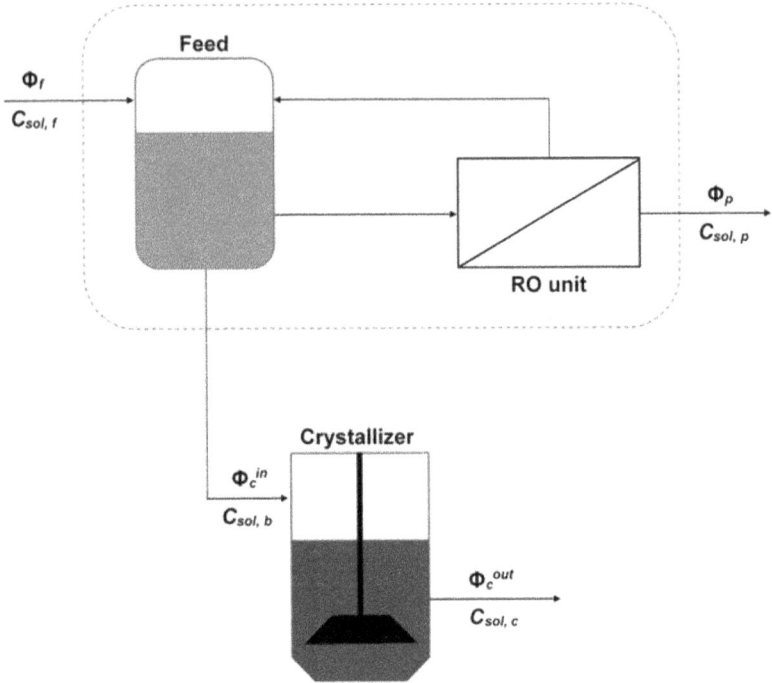

Figure 1.4.   Schematic representation of the RO-assisted crystallization setup.

membrane operations. It is referred to as RO-based MAC; to date, it is the most used membrane operation among pressure-driven unit.

1.5.1 *Mass balances*

Figure 1.4 provides a schematic representation of a RO-assisted crystallization unit.[39] Because the flow Φ to from the membrane is very large compared to the feed flow to the crystallizer, the feed vessel and the RO membrane unit are considered as a single well-mixed unit (dashed box in the figure).

The accumulation of total mass, M_b, in the feed/membrane vessel under negligible evaporation is

$$\frac{dM_b}{dt} = \Phi_f - \Phi_p - \Phi_c^{in}. \tag{1.18}$$

The accumulation of solute in the feed/membrane vessel is

$$\frac{dM_{Sol,b}}{dt} = \Phi_f C_{Sol,f} - \Phi_p C_{Sol,p} - \Phi_c^{in} C_{Sol,b}. \tag{1.19}$$

The solute mass fraction in the feed/membrane vessel $C_{Sol,b}$ is defined as

$$C_{Sol,b} = \frac{M_{Sol,b}}{M_b}. \tag{1.20}$$

It is assumed that crystals are produced and retained in the crystallizer, so that there is no flow of crystals to from the crystallizer, the crystals have 100% purity, and evaporation is negligible. The accumulation of total mass in the crystallizer is

$$\frac{dM_c}{dt} = \Phi_c^{in} - \Phi_c^{out}. \tag{1.21}$$

The accumulation of aqueous mass in the crystallizer is

$$\frac{dM_c^{aq}}{dt} = \Phi_c^{in} - \Phi_c^{out} - p_{Sol,c}, \tag{1.22}$$

where $p_{Sol,c}$ is the crystal production rate in the crystallizer. The accumulation of aqueous solute in the crystallizer is

$$\frac{dM_{Sol,c}^{aq}}{dt} = \Phi_c^{in} C_{Sol,b} - \Phi_c^{out} C_{Sol,c} - p_{Sol,c}, \tag{1.23}$$

where the aqueous solute mass fraction $C_{Sol,c}$ is defined as

$$C_{Sol,c} = \frac{M_{Sol,c}^{aq}}{M_c^{aq}}. \tag{1.24}$$

1.5.2 Membrane transport

For selective water (w) permeation, the chemical potential difference $\Delta\mu_w$ across the membrane can be related to the pressure difference by[40,41]

$$-\frac{\Delta\mu_w}{RT} = \ln\left(\frac{x_w^p \gamma_w^p}{x_w^m \gamma_w^m}\right) + \frac{1}{RT}\int_{P^m}^{P^p} v_{0,w} dP, \tag{1.25}$$

where x_w^p and x_w^m are the water molar fractions in the permeate (p) and at the membrane (m), γ_w^p and γ_w^m are the activity coefficients in the permeate and at the membrane, R is the gas constant, T is the absolute temperature, P is the pressure, and $v_{0,w}$ is the water partial molar volume at the permeate side. At equilibrium, the pressure dependency of $v_{0,w}$ is neglected, the transmembrane (tm) osmotic pressure $\Delta\Pi_{tm}$ is calculated as the difference between the osmotic pressure at the membrane (on the retentate side), Π_r, and the osmotic pressure of the permeate, Π_p:[41]

$$\Delta\Pi_{tm} = \int_{\Pi^p}^{\Pi^m} dP = \frac{RT}{v_{0,w}} \ln(x_w^m \gamma_w^m). \tag{1.26}$$

At low solute concentrations, Eq. (1.26) can be approximated by the Van't Hoff equation,

$$\Delta \Pi_{tm} = \frac{RT}{v_{0,w}} x_{Sol}^m = \frac{RT \rho_{Sol}}{W_{Sol}} x_{Sol}^m, \tag{1.27}$$

where x_{Sol}^m is the molar fraction of solute at the membrane surface, ρ_{Sol} is the solute density, and W_{Sol} is the molar mass of the solute.

Assuming a linear relation between the water molar flux and the difference in chemical potential, and a Henry-type behavior of water within the membrane material, the pressure difference can be related to the water flux across the membrane, J_w, by the solution diffusion model.[41–43] Accordingly, the water flux across the membrane is dependent on the water permeability of the membrane, A_w^m, and the net transmembrane pressure, ΔP_{tm}, as follows:

$$J_w = A_w^m (\Delta P_{tm} - \Delta \Pi_{tm}) \approx \frac{K_w^m D_w^m}{RT \delta_m} (\Delta P - \Delta \Pi_{tm}). \tag{1.28}$$

The permeate is always at atmospheric pressure while the pressure drop within the module is negligible; therefore ΔP_{tm} is equal to the applied pressure ΔP, which is measured at the retentate line (P^r). Furthermore, considering Eq. (1.27), Eq. (1.28) can be rewritten as

$$J_w = A_w^m \left(P^r - \frac{RT \rho_{Sol}}{W_{Sol}} \left(x_{Sol}^m - x_{Sol}^p \right) \right). \tag{1.29}$$

Neglecting the contribution of the other solutes to be negligible, the flux of solute is controlled by diffusion only:

$$J_{Sol} = B_{Sol} \left(x_{Sol}^m - x_{Sol}^p \right), \tag{1.30}$$

where B_{Sol} is the mass transfer coefficient across the membrane.

The membrane water permeability term includes the membrane thickness, δ_m, and the solubility, K_w^m, and diffusivity, D_w^m, of the permeating component in the membrane. Both K_w^m and D_w^m are temperature-dependent parameters, which can be described by an Arrhenius-type equation,[44] where the temperature dependencies of the solubility and diffusivity are related to the heat of solution, $\Delta_S H_w^m$, and the activation energy of diffusion, E_A, respectively:

$$K_w^m = K_{w,0}^m \exp \left(\frac{-\Delta_S H_w^m}{RT} \right) \tag{1.31}$$

$$D_w^m = D_{w,0}^m \exp \left(\frac{-E_A}{RT} \right). \tag{1.32}$$

These temperature dependencies can be combined in an apparent activation energy, E_{app}, describing the temperature dependency of the permeance by

$$P_w^m = \frac{P_{w,0}^m}{T} \exp\left(\frac{-E_{app}}{RT}\right). \quad (1.33)$$

Substituting Eq. (1.33) into Eq. (1.27) gives

$$\ln\left(\frac{J_w T}{\Delta P - \Delta \Pi_{tm}}\right) = \ln(P_{w,0}^m) + \left(\frac{-E_{app}}{R}\right)\frac{1}{T}. \quad (1.34)$$

The permeance also depends on the solvent feed concentration. This is a complex relation, which involves the interaction of the solute with the membrane material, and the influence on the water solubility. The total concentration adsorbed in the membrane can be assumed to be constant and described by a Langmuir isotherm.[45] The adsorbed solute can partly block the membrane, decreasing the effective surface area for water transport and, ultimately the overall permeability. The concentration dependency of $P_{w,0}^m$ can then be written as[46]

$$P_{w,0}^{m,eff} = P_{w,0}^m \left(\frac{1}{1 + K_0 C_{Sol}}\right). \quad (1.35)$$

1.5.3 Concentration polarization

Concentration polarization indicates the increase of the concentration of retained components at the membrane surface (C_{Sol}^m) compared to the bulk concentration (C_{Sol}^r), reducing the driving force to permeation. Since the crystallization process is operated close to saturation, an increase in concentration can induce scaling. Promoting turbulent flow at the retentate side of the membrane module can reduce polarization effects.[47]

For the water and solute flux in the liquid boundary layer, the Maxwell–Stefan equations in a simplified form can be used:[48]

$$\frac{dx_w}{dz} = \frac{x_{Sol} J_w - x_w J_{Sol}}{C D_w} \quad (1.36)$$

$$\frac{dx_{Sol}}{dz} = \frac{x_w J_{Sol} - x_{Sol} J_w}{C D_{Sol}^w}, \quad (1.37)$$

where z is the length-coordinate in the direction of the transport, J is the transmembrane flux, x is the molar fraction and C is the total concentration in the system which, at low solute concentrations, can be assumed to be equal to the molar density of the solvent (water). The diffusivity of solute in water is denoted by D_{Sol}^w, while D_w is the self-diffusion coefficient of water, which can be used because of the low solubility of the solute. Due to polarization effects, there is an additional mass transfer resistance in the boundary layer at the retentate side of the membrane. For a highly selective membrane ($J_{Sol} = 0$), the concentration gradient of solute in the boundary layer can be found by integrating Eq. (1.37) over the linearized boundary layer thickness, δ. The mass transfer of solute across the polarized layer is given by[43]

$$x_{Sol}^m - x_{Sol}^p = (x_{Sol}^r - x_{Sol}^p) \exp\left(\frac{J_w \delta}{CD_{Sol}^w}\right). \tag{1.38}$$

Using Eqs. (1.27), (1.32), and (1.38), the water flux can be related to the pressure difference across the membrane and retentate side concentration:

$$J_w = \frac{P_{w,0}^{eff}}{T} \exp\left(\frac{-E_{app}}{RT}\right)(\Delta P - RT C_{Sol}^m). \tag{1.39}$$

The effective boundary layer thickness depends on the geometry of the membrane module and varies over the length of the membrane. In order to evaluate the boundary layer thickness for a turbulent fluid flowing in a pipe, the Chilton–Colburn analogy, can be applied to the Dittus–Boelter correlation and,[49]

$$Sh = 0.023 Re^{\frac{4}{5}} Sc^{\frac{1}{3}}, \tag{1.40}$$

where the Reynolds ($Re = \upsilon \rho d / \eta$) and Schmidt ($Sc = \eta / (\rho D)$) dimensionless numbers are related to the Sherwood number, defined as

$$Sh = \frac{\frac{k}{\delta} l_{geo}}{k} = \frac{l_{geo}}{\delta}. \tag{1.41}$$

To estimate the geometric length, l_{geo}, the radius of the membrane tube can be used. Since the solute concentration C_{Sol} is equal to Cx_{Sol}, an expression for the solute concentration at the membrane surface, C_{Sol}^m, can be found:

$$C_{Sol}^m = C_{Sol}^r \exp\left(\frac{J_w}{k_{Sol}}\right), \tag{1.42}$$

where k_{Sol} is the mass transfer coefficient in the polarized layer.

Equations (1.29) and (1.30) thus become

$$J_w = A_w \left(P^r - \frac{RT \rho_{Sol}}{W_{Sol}} (x_{Sol}^r - x_{Sol}^p) \exp\left(\frac{J_w}{k_{Sol}}\right)\right) \tag{1.43}$$

$$J_{Sol} = B_{Sol} (x_{Sol}^r - x_{Sol}^p) \exp\left(\frac{J_w}{k_{Sol}}\right). \tag{1.44}$$

The flux of solute is calculated from the water flux and the permeate composition:

$$J_{Sol} = J_w x_{Sol}^p. \tag{1.45}$$

The selectivity of the membrane is conventionally evaluated by the retention coefficient,[43]

$$R_{Sol} = 1 - \frac{x^p_{Sol}}{x^r_{Sol}}. \tag{1.46}$$

Combining Eqs. (1.44)–(1.46) and solving for R_{Sol} gives

$$R_{Sol} = \frac{J_w}{B_{Sol} \exp\left(\frac{J_w}{k_{Sol}}\right) + J_w}. \tag{1.47}$$

In Eq. (1.47), R_{Sol} only depends on the water flux J_w and the mass transfer coefficients B_{Sol} and k_{Sol}. Substitution of Eqs. (1.46) and (1.47) into Eq. (1.43) allows at evaluating P^r as a function of J_w. In this way it is possible to individuate the maximum allowable values of J_w, taking into account the limit of operation of the membrane.

For a given membrane, B_{Sol} is a constant. Furthermore, in the case that the cross-flow velocity is approximately constant for all experiments, k_{Sol} is assumed to be constant. The values of B_{Sol} and k_{Sol} can be obtained from a set of experiments, in which x^r_{Sol}, x^p_{Sol}, and R_{Sol} are determined at various values of J_w, by using Eq. (1.47) to fit data points. The water permeability A_w can be derived from pure water flux measurements, for which Eq. (1.43) becomes

$$J_w = A_w P^r. \tag{1.48}$$

1.5.4 Steady-state equations

A constant crystal production rate can be achieved by keeping $C_{Sol,c}$ constant at steady state:

$$\frac{dC_{Sol,c}}{dt} = 0. \tag{1.49}$$

Differentiating Eq. (1.49) and solving for $p_{Sol,c}$ gives

$$p_{Sol,c} = \Phi^{in}_c \frac{C_{Sol,b} - C_{Sol,c}}{1 - C_{Sol,c}}. \tag{1.50}$$

In the feed vessel, $C_{Sol,b}$ and M_b should also remain constant:

$$\frac{dM_b}{dt} = 0, \tag{1.51}$$

$$\frac{dM_{Sol,b}}{dt} = 0. \tag{1.52}$$

Equations (1.18) and (1.19) thus become

$$\Phi_f = \Phi_p + \Phi_c, \tag{1.53}$$

$$\Phi_f C_{Sol,f} = \Phi_p C_{Sol,p} + \Phi_c C_{Sol,b}. \tag{1.54}$$

The permeate flow can also be expressed as

$$\Phi_p = J_W A, \tag{1.55}$$

where A is the membrane area. Substituting Eqs. (1.55) and (1.53) into Eq. (1.54) and solving for Φ_c gives

$$\Phi_c = J_W A \frac{C_{Sol,f} - C_{Sol,p}}{C_{Sol,b} - C_{Sol,f}}. \tag{1.56}$$

The concentration factor in the feed/membrane vessel is defined as

$$F = \frac{C_{Sol,b}}{C_{Sol,f}}. \tag{1.57}$$

Substituting Eqs. (1.46) and (1.57) into Eq. (1.56) gives

$$\Phi_c = J_W A \frac{1 - (1 - P_{Sol}) F}{F - 1}. \tag{1.58}$$

For a target crystal production rate $p_{Sol,c}$, Eq. (1.58) is substituted in Eq (1.50) to evaluate $C_{Sol,c}$ as a function of J_w. From Eq. (1.50) it can be seen that for $p_{Sol,c}$ to be large, $\Phi_{c,in}$ should be large, $C_{Sol,b} - C_{Sol,c}$ should be large, or $1 - C_{Sol,c}$ should be small.

1.5.5 Crystallization equations

For a seeded batch crystallization, under the assumptions of negligible nucleation, breakage, agglomeration, size-independent growth, and narrow size distribution of the seeds, the mass balance for the crystals becomes

$$p_{Sol,c} = -\frac{d M_{Sol,c}}{dt} = n_{Sol,c,0} \frac{dm_{Sol,c}}{dt}, \tag{1.59}$$

where $n_{Sol,c,0}$ is the number of seed crystals and $m_{Sol,c}$ is the mass of a single crystal. Assuming that the growth rate is proportional along the three dimensions, the aspect ratios remain constant. Therefore the crystal volume can be expressed with respect to of one of the dimensions, L, using the aspect ratio along the other dimensions. $n_{Sol,c,0}$ and $m_{Sol,c,0}$ can be defined as

$$n_{Sol,c,0} = \frac{M^s_{Sol,c,0}}{V^s_{Sol,0}\rho^s_{Sol}} = \frac{M^s_{Sol,c,0}}{L_0^3(AR_1)(AR_2)\rho^s_{Sol}}, \tag{1.60}$$

$$m_{Sol} = V^s_{Sol}\rho^s_{Sol} = L^3(AR_1)(AR_2)\rho^s_{Sol}, \tag{1.61}$$

where $M^s_{Sol,c,0}$ is the initial mass fraction of crystals and ρ^s_{Sol} is the density of the crystals. Equation (1.59) thus becomes

$$p_{Sol,c} = \frac{3M^s_{Sol,c,0}}{L_0^3} \frac{dL}{dt}. \tag{1.62}$$

From Eq. (1.60) it also follows that

$$\left(\frac{L}{L_0}\right)^2 = \left(\frac{M^s_{Sol,c,0}}{M^s_{Sol,c,0}}\right)^{\frac{2}{3}}. \tag{1.63}$$

The growth rate dL/dt can be expressed as a function of the supersaturation S:

$$\frac{dL}{dt} = k_g(S-1)^g = k_g\left(\frac{C_{Sol,c}}{C_{Sol,c,eq}} - 1\right)^g = k_g\left(\frac{M^{aq}_{Sol,c}}{M^{aq}_c C^{aq}_{Sol,c,eq}} - 1\right)^g, \tag{1.64}$$

where $C_{Sol,c,eq}$ is the mass fraction of solute at equilibrium (i.e. the solubility). Substitution of Eqs. (1.63) and (1.64) into Eq. (1.62) yields

$$p_{Sol,c} = 3(M^s_{Sol,c})^{\frac{2}{3}} \frac{(M^s_{Sol,c,0})^{1/3}}{L_0} k_g\left(\frac{M^{aq}_{Sol,c}}{M^{aq}_c C^{aq}_{Sol,c,eq}} - 1\right)^g. \tag{1.65}$$

The kinetic parameters k_g and g can be obtained from batch experiments by recording the mass of solute in the solution in time and fitting the data to Eqs. (1.43) and (1.65).

1.5.6 Energetic aspects

The energy requirement for RO-MAC is the sum of the electrical power for the pump,

$$E_{Pump} = \frac{F_{C,Out}\Delta P}{\eta}, \tag{1.66}$$

and the duty of the heat exchanger for the inlet stream of the membrane module,

$$E_{Hex} = F_{C,Out}\rho_L C_{PL}(T_m - T_C),$$ (1.67)

where η is the pump efficiency, ΔP is the pressure gradient, ρ_L is the liquid density, C_{PL} is the heat capacity liquid phase, T_m is the temperature at the membrane surface, T_C is the temperature in crystallization vessel, and $F_{C,Out}$ is the crystal-free volumetric liquid flow rate taken out of the crystallizer. For conventional evaporative crystallization or MD-MAC, the total energy requirement is given by

$$E_{Evap} = \frac{1 - w_F}{w_F} F_S \rho_S \Delta H_{Vap},$$ (1.68)

where w_F the weight fraction at the feed side of membrane (bulk), ρ_S the solid phase density, ΔH_{Vap} the heat of vaporization, and F_S the solid volumetric flow rate out of the system.

By combining Eqs. (1.66)–(1.68), an expression for the ratio of the energy requirement for RO-MAC over the energy requirement for the single-step evaporative crystallization process can be found as[50]

$$\xi_{RO/Evap} = \frac{E_{Pump} + E_{RO,Hex}}{E_{Evap}}$$

$$= \frac{[C_{PL}\rho_L\eta(T_m - T_C) + \Delta P]w_{m,Out}}{\rho_L\eta\{w_{Sat,T_C}[C_{PL}(T_m - T_C) - \Delta H_{Vap}] + w_{m,Out}\Delta H_{Vap}\}}.$$ (1.69)

The solubility behavior of a solute water system is assumed to be described by

$$w_{Sat}(T_m) = w_{Sat}(T_C)\exp[a\,(T_m - T_C)].$$ (1.70)

Equation (1.70) contains two parameters: the solubility at the minimum temperature in the system $w_{Sat}(T_C)$ and the exponential factor a, which is a measure of the steepness of the solubility curve. Both parameters need to be obtained from experimental solubility data. Furthermore, the system is constrained by a maximum pressure at which membranes can be operated. The polymeric membrane materials also exhibit limited temperature stability.

The minimum value of $\xi_{RO/Evap}$ for a given system is obtained by solving the following optimization problem:

$$\min_{T_m} \xi_{RO/Evap}(T_m)$$ (1.71)

$$w_{m,Out} = w_{Sat}(T_m)$$ (1.72)

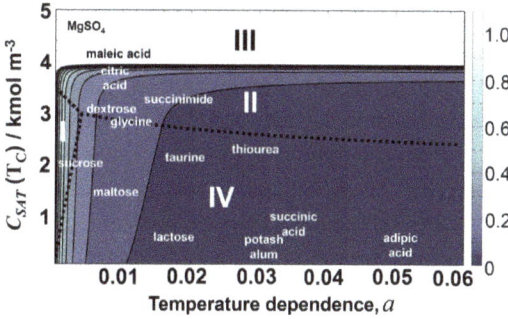

Figure 1.5. Contour plot describing the development of the minimum value of $\xi_{RO/EVAP}$ as a function of solubility parameters C_{SAT} (T_C) and a. Concentration is expressed on a molar basis. The dotted lines represent boundaries between the regions with different constraints. The approximate locations of typical components in the diagram are obtained for reference by fitting solubility data from literature to Eq. (1.70) [Reprinted from [50], Copyright (2014), with permission from Elsevier].

$$\Delta P = \frac{T_m \rho_L w_{m,Out} R}{M_{W,A}} < \Delta P_{Max} \qquad (1.73)$$

$$T_C \leq T_m \leq T_{Max}. \qquad (1.74)$$

Figure 1.5 illustrates the potential energy savings that can be achieved by generating supersaturation for crystallization with RO membranes compared to single-step evaporative crystallization (both conventional or MD-MAC) as a function of the solubility parameters. The diagram can be divided into four regions:

1. A region I constrained by the maximum temperature in the membrane module T_{MAX}. This region contains systems with a flat solubility curve;
2. A region II constrained by the maximum osmotic pressure ΔP_{MAX} at the outlet of the membrane module. For this region, the osmotic pressure of the system at the conditions in the crystallizer is close to the maximum allowed pressure in the membrane module. A steeper solubility curve widens this region;
3. An infeasible region III for RO-MAC. In this region, the osmotic pressure at the conditions in the crystallization vessel exceeds the maximum osmotic pressure ΔP_{MAX}. Both conventional evaporative crystallization and MD-MAC are feasible alternatives for components in this region. Also, the osmotic pressure at the permeate side of the membrane can be increased to make the process feasible;
4. An unconstrained region IV;

It can be concluded from Fig. 1.5 that a system with a steep solubility curve has a lower optimal temperature in the membrane module and, consequently, a lower energy requirement for RO-MAC compared to systems with a flat solubility curve. Within the feasible domain of RO-MAC there is a narrow range for which evaporative crystallization is energetically more favorable, which is either a system with an osmotic pressure in the crystallization vessel close to the maximum allowed pressure, or systems with a very flat solubility curve. This region can be increased by applying multi-stage evaporation at the expense of expected increased capital costs. For systems with a large molecular weight (low molar concentration), RO-MAC has larger feasibility region and, consequently, a larger region in which the energy requirements are significantly less compared to evaporative crystallization. There is a sharp distinction between the two regions in which evaporative crystallization or RO-MAC is more favorable. That is, by crossing the boundary between these two regions either RO-MAC becomes infeasible very quickly or the benefit of RO-MAC compared to evaporative crystallization rises sharply. This analysis shows that both MD-MAC and RO-MAC technologies are complementary in different conditions on the energetic point of view.

1.6 Membrane Distillation (MD) and Osmotic Distillation (OD)

MD is a thermally-driven process in which a hydrophobic porous membrane separates warm and cold solution streams.[51,52] The hydrophobic nature of the membrane prevents the passage of liquids through the pores, while allowing the passage of volatile solvent in vapor phase (Fig. 1.6). The temperature gradient across the two sides of the membrane produces a vapor pressure gradient, which causes solvent vapor to diffuse through the membrane and to condense on the colder solution surface. Therefore, the transport mechanism is broken down into three steps: evaporation at the membrane interface at the feed side, migration of the vaporized solvent through the dry membrane pores, and condensation on the opposite distillate side.[22] The result is a distillate of very high purity which, unlike in conventional distillation, does not suffer from nonvolatile contaminants.

OD is a concentration technique for aqueous mixtures in which a volatile component (commonly water) is removed from the feed side by using a hypertonic draw solution flowing downstream a porous hydrophobic membrane. Vapor diffuses through the membrane pores under a driving force, given by the vapor pressure difference between both membrane sides and generated by the different activities of the feed and permeate streams. OD is not an athermal mass transfer operation; since transport involves solvent evaporation at the feed side and its condensation at the stripping side. A temperature difference at the membrane interfaces is thus

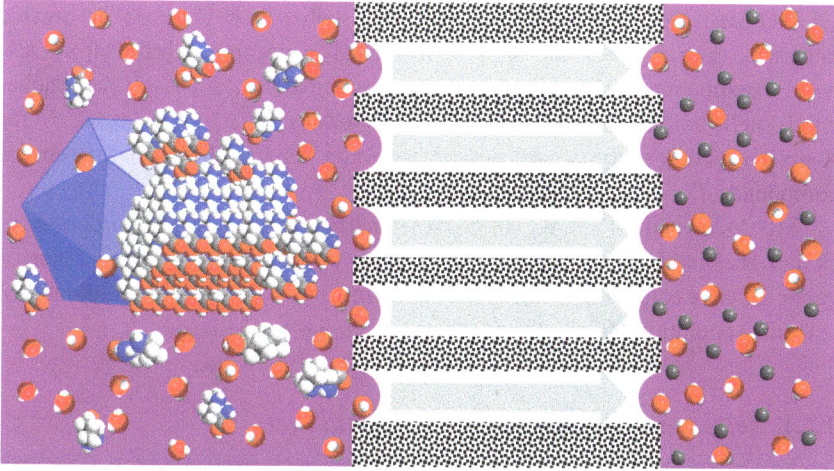

Figure 1.6. Working principle of a membrane crystallizer. The aqueous feed solution contains the substance to be crystallized on the left side of the porous hydrophobic membrane. Vapor pressure gradient between the two membrane sides drives the transport mechanism consisting into three steps: evaporation at the membrane interface at the feed, migration of the vaporized solvent through the dry pores, and condensation on the opposite distillate side.

created, even if the bulk temperatures of the two liquids are equal. In general, the temperature difference in aqueous systems is lower than 1°C, leading to a negligible decrease of the vapor flux. OD-MAC can be therefore used for the crystallization of molecular systems that are particularly sensitive to heat, such as biological macro-molecules, since solvent evaporation is not driven by the heating of the crystallizing solution, as in conventional thermal evaporators or in MD-MAC.

In both MD- and OD-MAC, hydrophobic porous membranes are used to maintain tight control of supersaturation while crystallization takes place in a circulating crystallizer.[22,53] The membrane surface does not act simply as a support for solvent evaporation but it is also be active to stimulate heterogeneous nucleation, so that crystallization will start at lower supersaturation ratios than in the homogeneous case.

The MD-MAC/OD-MAC method provides an experimental setup for controlling the kinetic parameters in crystallization process, to produce controlled-size, well-shaped, and pure crystals; in the case of proteins, the low level of impurities inclusions in the crystalline lattice is associated to high diffracting power. This is because the rate of solvent removal and the rate of supersaturation generation can be controlled by selecting the opportune membrane physico-chemical properties and the proper operating conditions.

MD-MAC/OD-MAC have been used to crystallize both inorganic and organic materials, such as NaCl and MgSO$_4$ from high concentrated brines from seawater desalination,[11,22] for the preparation of fumaric acid crystals from aqueous L-malic acid solutions,[54] proteins like lysozyme[15,16,19] and trypsin[17,55] for X-ray diffraction analysis purposes, and selected polymorphs of pharmaceutical compounds.[56]

MD-MAC and OD-MAC are conventionally classified as MCr processes,[22] as they are both based on the same driving force for evaporation (the gradient of vapor pressure across the membrane). MCr can be operated in two configurations: solvent evaporation MCr and antisolvent membrane crystallization.

1.6.1 *Solvent evaporation membrane crystallizer*

In this configuration, the crystallizing solution is located at the feed side of the membrane, as depicted in Fig. 1.6. In accordance with the general working principle of a membrane crystallizer, described in the previous section, the membrane must be prevented from being wetted by the contacted solutions, because mass exchange has to occur only in vapor phase. To avoid wetting, hydrophobic membrane materials are preferred for aqueous solutions; based on the same principle, oleophobic membranes can be used for organophilic liquids in crystallization from organic solvents. In the most common arrangement, the distillate side of the membrane consists of a condensing fluid at a lower temperature than the feed side (often the pure solvent), in the case of MD-MAC, or in a draw hypertonic solution of inert salts (e.g. NaCl, CaCl$_2$, etc.) in the case of OD-MAC. As solvent evaporates from the feed, the establishment of the concentration polarization layers adjacent to the membrane face occurs. Accordingly, the solute concentration in the polarized layer close to the membrane surface, c_m, is higher with respect to the bulk solute concentration, c_b, on the feed sides. According to Eq. (1.73), this concentration polarization layer is related to the transmembrane flux J:[57]

$$\frac{c_m}{c_b} = \exp\left(\frac{J}{k}\right),\qquad(1.75)$$

where k is the mass transfer coefficient. In MCr low fluxes (in the order of few L·h^{-1}·m^{-2}) are usually used to control the supersaturation. Therefore, concentration polarization does not represent a limiting factor for the driving force, as it is instead in RO. Due to the change of the physical state of the volatile component (solvent) at the interface between the solutions and the membrane, heat is absorbed by vaporized molecules on the feed side, while it is released on the distillate side upon condensation, so that temperature polarization establishes across the membrane, thereby lowering the net driving force for the process. Although temperature polarization generally represents the major limiting factor in MD/OD,[51] in the case of MCr both concentration and temperature polarization might affect locally the degree of supersaturation so that nucleation can proceed under a different mechanism close to the membrane with respect to the bulk. The crystals nucleated and grown on/near the membrane might therefore display peculiar characteristics due to this phenomenon.[58]

1.6.2 *Antisolvent membrane crystallizer*

Antisolvent membrane crystallizers operate according to the two schemes depicted in Fig. 1.7 In solvent/antisolvent demixing configuration (Fig. 1.7 up), a certain solute is dissolved in an appropriate mix of a solvent and an antisolvent. When a gradient of vapor pressure is generated between the two sides of the membranes by a temperature difference, the solvent, which has a higher vapor pressure than the antisolvent at the same temperature, evaporates at a higher rate, thus producing a certain degree of solvent/antisolvent demixing. As the fraction of antisolvent in the mixture increases from the initial composition, up to a certain system-specific threshold value, the lower solubility of the solute generates supersaturation and a phase separation occurs. The requirements for this configuration are that: (i) the antisolvent and the solvent are miscible; (ii) the solute is under its solubility limit in the initial solvent/antisolvent mixture; (iii) the solvent evaporates at higher velocity than the antisolvent; (iv) the solvent and the antisolvent do not form azeotropes or, if they do, supersaturation occurs before the azeotropic composition point. This configuration might be particularly useful for the crystallization of species recovered by aqueous/organic solvent extraction combination mixtures. A typical example is an organic molecule for which water is the antisolvent and an alcohol is the solvent and that is recovered by liquid–liquid extraction methods.

In antisolvent addition configuration (Fig. 1.7), a solute is dissolved in a solvent and, then, an antisolvent is gradually evaporated from the other side of the membrane by applying a gradient of vapor pressure. As the antisolvent mixes with the solvent in the feed solution, solute solubility in the solvent/antisolvent mixture reduces, so that supersaturation and solute crystallization is obtained when the fraction of antisolvent exceeds a certain system-specific threshold value. This configuration requires that the antisolvent and the solvent are miscible. In this case, the species to be crystallized can be soluble in aqueous solutions and poorly soluble in organic, low-boiling liquids.

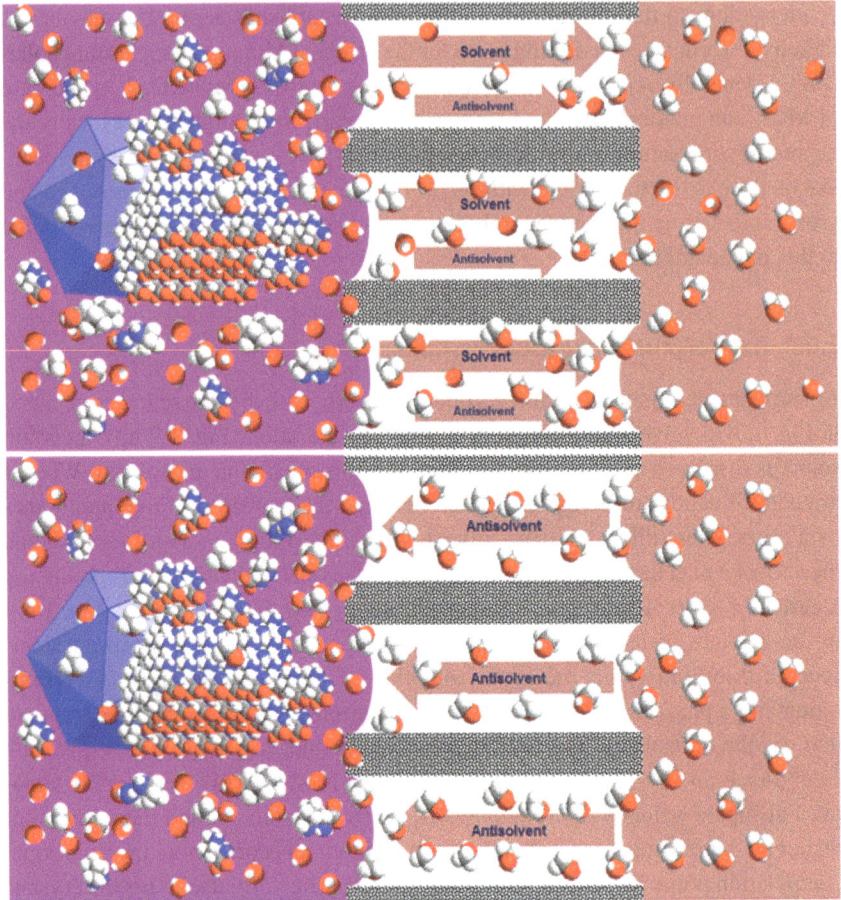

Figure 1.7. Working principle of antisolvent MCr. Up: the substance to be crystallized is recovered in a solvent/antisolvent mixture. The solvent, having a high vapor pressure than the antisolvent, flows at higher rate through the membrane, so that the shift in mixture composition allows the reduction in solute solubility. Down: an antisolvent is added in vapor phase through the membrane in a solution containing the species to be crystallized. As the antisolvent diffuses in the solutions, solubility of solute decreases and, as a consequence, crystallization occurs.

1.6.3 *General advantages of MCr*

The general potentialities of MCr are related to the intrinsic properties of membrane processes, well responding to the requirements of Process Intensification. More specifically, when using MCr technology, the interfacial area for solvent removal or antisolvent addition per unit of crystallizer volume is constant and very high — especially in the case of hollow fibers membrane modules (more than $20.000 \, m^2/m^3$,[59]) — while the removal/addition points are distributed over the whole extension of the membrane (pores). Accordingly, supersaturation gradients can be substantially reduced when designing a crystallizer. This possibility is hardly accessible in the case of conventional thermal evaporators.

MCr based on MD and OD differs from other MAC processes because mass transfer sustained by the membrane does not occur by size exclusion effect, but by a mechanism of evaporation-diffusion-condensation. This mechanism, occurring in vapor phase, allows a finer control over the transmembrane flux and, hereafter, on supersaturation. Low-pressure MF membranes, with nominal pore sizes in the order of a few hundreds of nanometers, are utilizable for MCr applications. Therefore, non-volatile solute molecules, with a molecular size much smaller than the pores' size, can be crystallized while keeping suitable transmembrane fluxes (productivities) associated to very low pumping energy input (the only required if for the recirculation of solutions close to atmospheric pressure).

Flux decline with increased feed concentration is an inherent characteristic of MD. As the concentration of the solute increases, the vapor pressure of the solvent decreases according to Raoult's law, and the driving force for solvent removal — the partial pressure difference — is reduced. Concentration and temperature polarization, as well as lowering the vapor pressure associated with an increase in solute concentration, might cause a gradual decrease in flux, up to a critical supersaturation. A drastic decline in flux beyond this critical supersaturation is due to a rapid increase in crystal deposition on the membrane and loss of membrane permeability.[22,60] On the other side, in the case of MCr, the establishment of the conditions suitable for crystallization (supersaturation generation) sustained by the membrane and, at the same time, the crystallization phenomenon which withdraws small particles (crystals) from the solution that might foul the membrane, represent a particular synergic combination, useful to increase the membrane life time.

Useful even in the case of heterogeneous nucleation induced by the membrane surface, the shear stress generated in a dynamic membrane crystallizer is generally sufficient to avoid crystals deposits, thus preserving the membrane functionality. However, if considering working operation over a long time, membrane cleaning would surely be advisable, as for every membrane operation.

A fundamental point relies on the energetic penalty involved in MCr operations. In this process, the evaporation step requires an obligatory energy input, consisting of the latent heat of vaporization associated with the phase changes.[51] This energy contribution makes MCr energetically comparable with evaporator crystallizers. In this respect, RO-MAC, although requiring high pumping energy to generate usually 40–60 bar, become economically more feasible with respect to thermal evaporation at least for substances with moderate solubility, larger molecular weight, and a strong dependency of their solubility on temperature.[50,61] On the other side, RO-MAC is limited by concentration polarization phenomena, which causes reduction of the driving force for mass transport across the membrane and, eventually, solute precipitation above the membrane, with consequent loss of selectivity. For highly soluble and low molecular weight substances, the osmotic pressure of the solution, which has to be overcome to achieve positive and economically suitable transmembrane fluxes, is so high that the pumping energy required makes the process energetically unfavorable; furthermore, the required high pressure exceeds the maximum allowable limit for conventional membrane modules, not preserving module integrity. Nevertheless, substances displaying flat solubility curves require such a high amount of heat to warm the solution before the RO stage, to avoid scaling above the membrane, that the heat consumption makes the process economically unfeasible. Dynamic MCr does not generally suffer from the limitations linked to the concentration polarization or to the osmotic pressure. Temperature polarization, which usually limits MD operations, is not so critical in MCr, as a temperature as high as 30–40°C is enough to produce the transmembrane flux (comparable with RO-MAC) suitable to generate supersaturation.[62,63] Nevertheless, if the development of new membranes with reduced thermal conductivity would allow to reduce further the heat lost through the membrane and thermal polarization phenomena, the energetic gap between MCr with respect to RO-MAC might be covered by using waste heat when the process is driven by e.g. solar energy or low-grade heat sources,[64–66] thus increasing substantially the overall energetic efficiency of the process.

The main concerns about MCr are related to the membrane stability. In the case of membrane wetting, due to the loss of hydrophobicity, the process completely collapses because of the direct dispersion of one phase into the other without any selectivity. This potential risk can be reduced by operating with moderate transmembrane fluxes (in the order few L/m^2·h) and sufficiently hydrophobic membrane materials (like polypropylene, polytetrafluoroethylene, polyvinylidene fluoride).

1.7 Other Membrane-Assisted Crystallization Configurations

1.7.1 *Membrane contactors (MCs)*

MCs involve one phase introduced through the membrane pores coming into contact with a second phase that flows tangentially to the membrane surface or in a frontal configuration. MCs were recently investigated for precipitation and crystallization operations.

Precipitation typically consists of mixing two liquid streams to create supersaturation, which then induces nucleation, particle growth, and agglomeration. The hydrodynamic factors governing the initial contact and mixing of the reacting fluids play a major role in controlling particle quality, due to their influence on the supersaturation distribution in the reactor volume.[67,68] Most of the traditional crystallizers, such as stirred tank reactors, operate with some degree of mixing generated by internal impellers or by pumping. However, these configurations produce high complexity in their internal hydrodynamics, particularly the distribution of the local power dissipation rate, influencing the intensity of mixing, or they can be obstructed by solid particles, either bound to or nucleated on the inner surfaces.[69]

MCs provide interesting alternatives to conventional micro-mixer geometries.[70,71] MCs based on porous hollow fiber devices are able to provide high supersaturation levels. Moreover, supersaturation is generated uniformly, due to the large number of feed introduction points — the membrane pores. In addition, radial mixing is substantial, in contrast to traditional tubular devices, and the typical time required in this process is comparable to the device residence time.[27]

An additional advantage of the membrane contactor is the easy scale-up.[27,72] Flow rates can easily be increased by adding more fibers to the device and by using devices in parallel, thereby achieving the desired conversion rate and productivity. A further potential advantage of these membrane-based crystallization processes is their higher capacity to generate the required number of nuclei/crystals per unit volume. Large quantities of particles can be produced in each product run; therefore, the efficiency of the process may be very high. Average particle size can also be modified by an appropriate choice of process parameters (e.g. membrane pore size, cross-flow velocity, pressure, formulations, etc.).

1.7.2 *Membrane reactors (MRs)*

In these systems, a solution containing the reagent A in pressed through a porous membrane in the solution of the reactant B. As a result, the solid product AB precipitates if the micro-mixed solution is supersatured with respect to it (Fig. 1.8).

Barium sulfate ($BaSO_4$) and calcium carbonate ($CaCO_3$) nanoparticles were successfully obtained using hollow fiber membranes with various MWCO.[29,73,74] Nano-sized $BaSO_4$ particles, about 15 nm in size, were synthesized in a MR in which Na_2SO_4 solutions were added gradually to the $BaCl_2$ solutions through

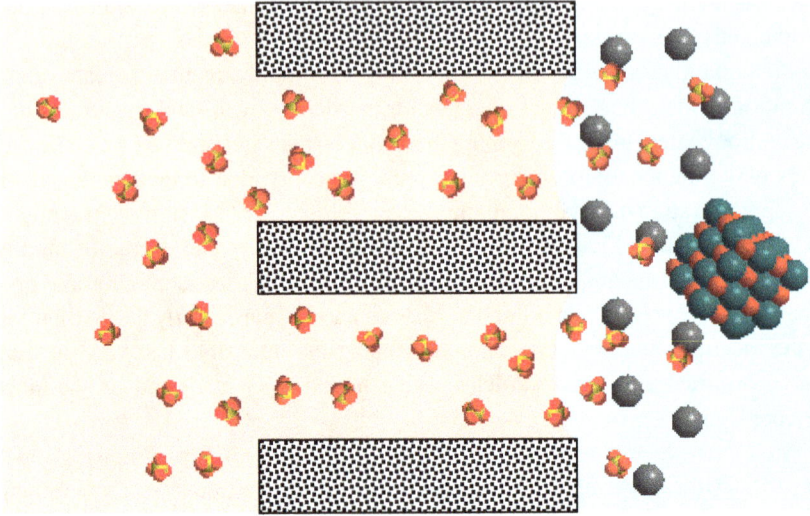

Figure 1.8. Working principle of the MRs-MAC process. A solution of component A flows tangentially to the membrane surface and reacts with a solution of component B introduced through the membrane pores. The supersaturation induces nucleation and particle growth.

the pores of ultrafiltration (UF) membranes to control the saturation ratio and subsequent nucleation and growth rates.[77]

In such membrane system, typically using UF or MF membranes, transmembrane flux J_B can be described by[57]

$$J_B = \frac{\Delta P}{(R_m + R_f)\mu},\tag{1.76}$$

where ΔP is the transmembrane pressure, R_m the resistance of the membrane, R_f the resistance of the fouling layer and concentration polarization, and μ the viscosity of solutions, which can be expressed as[75]

$$\mu = \frac{Nh}{V}\exp\left(\frac{3.8T_b}{T}\right),\tag{1.77}$$

where N is the Avogadro constant, h is Planck's constant, V is the molar volume, and T_b is the boiling point of the solution. As supersaturation is directly related to the amount of reactant B permeating through the membrane, increasing flux (for example, by increasing ΔP), would enhance precipitation kinetics. Higher particle growth unit concentration in solution and aggregation growth rate lead to poor integration degree of the growth units. Particle aggregation also increases with membrane MWCO due to the transmembrane flux increase caused by the larger pores. With the increase in T, J_B increases because of the falling of viscosity of the solution. However, according to the Stokes–Einstein equation, the diffusion coefficient of growth units is expressed as

$$D = \frac{kT}{6\pi \mu r},\tag{1.78}$$

where r is the radius of the diffusion particles. As T rises, the diffusion coefficient of growth units towards the nucleus and the surface reaction rate increases, resulting in more regular nanoparticles. In the case of low T, generally only large irregular agglomerates (amorphous) can be obtained, which is attributed to the low diffusion coefficient and the low nucleus surface reaction rate. Under these conditions, the growth is most likely dominated by an aggregation growth mechanism (i.e. nuclei aggregate quickly because of their high surface energy) driven by weak forces, (e.g. van der Waals force, solvent force). The nuclei in contact are cemented together by the deposition of growth units from the bulk solution, and form amorphous particles. The aggregation rate is determined by the colloid stability factor, W, as well as the number density of the growth units in suspension. W is defined as follows:[76]

$$W = a + b \int_{a+b}^{\infty} \exp\left(\frac{V(R)}{kT}\right) \frac{dR}{R^2},\tag{1.79}$$

where a and b are the radius for two spheres separated by the distance of R, and $V(R)$ is the potential-energy function. Therefore, a higher temperature is preferable. Increased reactant concentration also has a direct effect on the increase of supersaturation, leading to loosely regular particles. Oppositely, when a membrane with low MWCO is used, under low transmembrane pressure, the lower permeation flux combined with the enhanced diffusion coefficient for high temperature and reduced reactant concentrations, leads to lower local saturation ratio, nucleation rate, and aggregation growth rate. The growth units are deposited on the aggregates gradually, resulting in higher integration degree and regular nanoparticles.

1.7.3 Antisolvent pressure-driven membrane-assisted crystallization

The configuration in which an antisolvent is added to the crystallizing solution using a porous membrane to produce particles with narrow CSDs was patented some years ago.[25] Compared to tubular precipitators, nucleation rates in the antisolvent hollow fiber crystallizers is several orders of magnitude higher at low supersaturation and comparable to — or slightly lower than — at larger supersaturation.[27] These high nucleation rates are attributed to the inherent characteristics of feed introduction that MAC devices offer. A large number of feed solution elements are created at the pore outlets, which are exposed to the high supersaturation levels created by their contact with the antisolvent. Similarly, high supersaturation can also be created in a stirred vessel. Comparatively, in a stirred vessel, mesomixing limitation would result in a lower number of nuclei and correspondingly larger particles.

The addition of antisolvent to a crystallizing solution flowing on the lumen side of a hollow fiber membrane, or the addition of the crystallizing solution on the lumen side to the antisolvent flowing over the shell side by keeping the lumen side pressure higher, are two possible configurations using hollow fibers.[27] Antisolvent crystallization of L-asparagine results in CSDs shifted to smaller sizes compared to batch-stirred crystallizers. A decrease in antisolvent flux is expected with the first configuration, because of possible fiber blockage due to crystals. Polyaluminum chloride nanoparticles,[74] an important flocculant in drinkable water and wastewater treatment, and Perylene nanoparticles[77] were synthesized by these MAC devices.

1.7.4 *Supported liquid membranes (SLMs)*

In SLMs, the membrane solvent, a mobile carrier, and a porous polymer membrane are the basic components. Initially, a component A (usually an ion) combines with the carriers on one side of the membrane, and then the complex is transported to the other side of the SLM, and thereby captured by a second (counter-ionic) component B, thus readily making the solution supersaturated. Then the supersaturated solution experiences the nucleation of the AB compound (Fig. 1.9).

The application of such technology has been focused on polymorphism and morphology control for several inorganic species, mainly carbonates:[78,79] calcium carbonate, strontium carbonate, $MnCO_3$, and other alkaline earth metal carbonates. Nanowires, rods, shuttles, and hollow spheres were synthesized in a SLM system, and different morphologies can be obtained by means of the cooperation between SLM and citric acid, nitrilotriacetic acid, and ethylenediaminetetraacetic acid, used as modifiers, which modulate the orientation of crystal growth.[80]

The SLM can mimic the behaviour of a biomembrane in three different ways: (i) the use of mobile carrier, which exists in both an SLM and a cell membrane; (ii) the growth of the crystal in selective position; (iii) the presence of both mineralization processes in the SLM system and natural biomineralization processes in the organic medium.

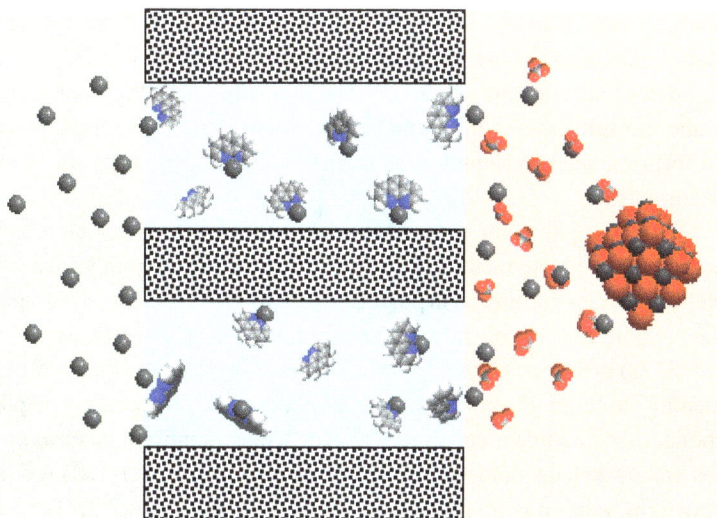

Figure 1.9. Schematics of MAC based on SLMs. Here, a carrier confined inside the membrane pores selectively binds to a species A; the carrier-A adduct diffuses through the membrane and, on the other side, is released in the component B solution, whereby it combines with component B to precipitate as solid product.

1.8 Membrane Templates (MTs)

In contrast to previous membrane techniques, in MTs-MAC there is no mass transfer across the membrane to control and limit the maximum level of supersaturation in the membrane template method. The membrane is used passively as a mold (hard template) to control product parameters such as size and shape. Nanochannels are filled with the desired material, such as precursor, polymer, metal, semiconductor, or carbon. Subsequently, the membrane may be removed to leave the desired nanotubes, nanorods, and nanowires that replicate the size and morphology of the template.[81–85] The dimensions and shape of the structures are defined by the membrane support. Various chemical compositions can be targeted for the structures, ranging from carbon, metal, and polymer to inorganic materials. Electrochemical and electroless depositions, chemical polymerization, sol–gel deposition, chemical vapor deposition, and polymer-derived ceramics are major chemical strategies employed.

Most of the work in template synthesis has entailed the use of two types of membrane: track-etch polymeric membranes and porous alumina membranes. Microporous and nanoporous polymeric filtration membranes are prepared by the track-etch method.[81] The resulting membranes contain cylindrical pores of uniform diameter, which are randomly distributed in the volume of the membrane. Porous alumina membranes are electrochemically prepared from an aluminum sheet in an appropriate aqueous solution.[84] The pore structure of a porous alumina membrane is a self-ordered hexagonal array of cells, with cylindrical pores of almost uniform diameter and length, which can be controlled by changing the experimental conditions.

Several applications concern to the crystallization of inorganic salts using the membrane template method. Potassium iodide crystals were grown in a matrix of etched ion tracks produced in polycarbonate foils.[86] Calcite was precipitated within the cylindrical pores of track-etch membranes via an amorphous calcium carbonate (ACC) precursor phase.[87,88] Macroporous crystals with sponge-like structures identical to those of sea urchin skeletal plates were prepared by templating with a sponge-like polymer membrane.[89,90] A polymer membrane taken as a replica of the sea urchin skeletal plate was used as a mold. A variety of crystals with sponge-like structures were thus prepared: calcite, $SrSO_4$, $SrCO_3$, $PbSO_4$, $BaCO_3$, $MgCO_3$, and NaCl.

The template approach is a versatile method for producing single crystals with unique shapes. It is generally used to provide a simple and cost-effective method that allows the complex topology of a template to be duplicated. The major advantage is that both dimensions and compositions of the particles can be easily controlled by varying the experimental conditions. The main limits of such template-directed synthesis are that post-synthesis removal of the template may damage crystals. Moreover, the use of a template limits the quantity of structures that can be produced in each product run. Therefore, the yield and efficiency of membrane template-directed synthesis may be not sufficient for industrial use.

References

1. Zeppezauer, M., Eklund, E. and Zeppezauer, E. (1968). Micro diffusion cells for the growth of single protein crystals by means of equilibrium dialysis, *Arch. Biochem. Biophys.*, **126**, 564–573.
2. Boerlage, S.F.E., Kennedy, M.D., Bremere, I. *et al.* (2000). Stable barium sulphate supersaturation in reverse osmosis, *J. Memb. Sci.*, **179**, 53–68.
3. Tomaszewska, M. (1993). Concentration of the extraction fluid from sulfuric acid treatment of phosphogypsum by membrane distillation, *J. Memb. Sci.*, **78**, 277–282.
4. Gryta, M. (2000). Concentration of saline wastewater from the production of heparin, *Desalination*, **129**, 35–44.
5. Gryta, M., Tomaszewska, M., Grzechulska, J. *et al.* (2001). Membrane distillation of NaCl solution containing natural organic matter, *J. Memb. Sci.*, **181**, 279–287.

6. Gryta, M., Grzechulska-Damszel, J., Markowska, A. *et al.* (2009). The influence of polypropylene degradation on the membrane wettability during membrane distillation, *J. Memb. Sci.*, **326**, 493–502.

7. Van de Lisdonk, C.A.C., Rietman, B.M., Heijman, S.G.J. *et al.* (2001). Prediction of supersaturation and monitoring of scaling in reverse osmosis and nanofiltration membrane systems, *Desalination*, **138**, 259–270.

8. Azoury, R., Garside, J. and Robertson, W.G. (1986). Crystallization processes using reverse osmosis, *J. Cryst. Growth*, **79**, 654–657.

9. Azoury, R., Garside, J. and Robertson W.G. (1986). Habit modifiers of calcium oxalate crystals precipitated in a reverse osmosis system, *J. Cryst. Growth*, **76**, 259–262.

10. Azoury, R., Robertson, W.G. and Garside, J. (1987). Generation of supersaturation using reverse osmosis, *Chem. Eng. Res. Des.*, **65**, 342–344.

11. Wu, Y. and Drioli, E. (1989). The behaviour of membrane distillation of concentrated aqueous solution, *Water Treatment*, **4**, 399–415.

12. Wu, Y., Kong, Y., Liu, J. *et al.* (1991). An experimental study on membrane distillation-crystallization for treating wastewater in taurine production, *Desalination*, **80**, 235–242.

13. Curcio, E., Criscuoli, A. and Drioli, E. (2000). *Proceedings of Euromembrane*, Hills of Jerusalem, Israel, 24–27 September, 2000.

14. Curcio, E., Di Profio, G. and Drioli, E. (2002). Membrane crystallization of macro-molecular solutions, *Desalination*, **145**, 173–177.

15. Curcio, E., Di Profio, G. and Drioli, E. (2003). A new membrane-based crystallization technique: tests on lysozyme, *J. Cryst. Growth*, **247**, 166–176.

16. Di Profio, G., Curcio, E., Cassetta, A. *et al.* (2003). Membrane crystallization of lysozyme: kinetic aspects, *J. Cryst. Growth*, **257**, 359–369.

17. Di Profio, G., Curcio, E. and Drioli, E. (2005). Trypsin crystallization by membrane-based techniques, *J. Struct. Biol.*, **150**, 41–49.

18. Zhang, X., El-Bourawi, M.S., Wei, K. *et al.* (2006). Precipitants and additives for membrane crystallization of lysozyme, *Biotechnol. J.*, **1**, 1302–1311.

19. Zhang, X., Zhang, P., Wei, K. *et al.* (2008). The study of continuous membrane crys-tallization onlysozyme, *Desalination*, **219**, 101–117.

20. Todd, P., Sikdar, S.K., Walker, C. *et al.* (1991). Application of osmotic dewatering to the controlled crystallization of biological macromolecules and organic compounds, *J. Cryst. Growth*, **110**, 283–292.

21. Sluys, J.T.M., Verdoes, D. and Hanemaaijer, J.H. (1996). Water treatment in a Membrane-Assisted Crystallizer (MAC), *Desalination*, **104**, 135–139.

22. Curcio, E., Criscuoli, A. and Drioli, E. (2001). Membrane crystallizers, *Ind. Eng. Chem. Res.*, **40**, 2679–2684.

23. Van der Gun, M.A. and Bruinsma, O.S.L. (2005). Crystallization and Nanofiltration, Partners in Minerals Processing, *Proc. 16th International symposium on industrial crystallization*, Dresden, Germany, 11–14 September, 2005.

24. Campbell, J., Peeva, L.G. and Livingston, A.G. (2014). Controlling crystallization via organic solvent nanofiltration: The influence of flux on griseofulvin crystallization, *Cryst. Growth Des.*, **14**, 2192–2200.

25. Bakker, W.J.W., Geertman, R.M., Reedijk, M.F. *et al.* (2004). Antisolvent Solidification Process, World Pat. Appl. WO2004096405A1.

26. Mayer, M.J.J., Demmer, R.L.M., Van Strien, C.J.G. *et al.* (2004). Process Involving the Use of Antisolvent Crystallisation, *World Pat. Appl.* WO2004096404A1.

27. Zarkadas, D.M. and Sirkar, K.K. (2006). Antisolvent crystallization in porous hollow fiber devices, *Chem. Eng. Sci.*, **61**, 5030–5048.

28. Di Profio, G., Stabile, C., Caridi, A. *et al.* (2009). Antisolvent membrane crystallization of pharmaceutical compounds, *J. Pharm. Sci.*, **98**, 4902–4913.

29. Jia, Z. and Liu, Z. (2002). Synthesis of nanosized $BaSO_4$ particles with a membrane reactor: effects of operating parameters on particles, *J. Membr. Sci.*, **209**, 153–161.

30. Kieffer, R., Mangin, D., Puel, F. *et al.* (2009). Precipitation of barium sulphate in a hollow fiber membrane contactor, part I: Investigation of particulate fouling, *Chem. Eng. Sci.*, **64**, 1759–1767.

31. Kieffer, R., Mangin, D., Puel, F. *et al.* (2009). Precipitation of barium sulphate in a hollow fiber membrane contactor, Part II: The influence of process parameters, *Chem. Eng. Sci.*, **64**, 1885–1891.

32. Wu, Q.-S., Sun, D.-M., Liu, H.-J. *et al.* (2004). Abnormal polymorph conversion of calcium carbonate and nano-self-assembly of vaterite by a supported liquid membrane system, *Cryst. Growth Des.*, **4**, 717–720.

33. Zeelen, J.P. and Wierenga, R.K. (1992) The growth of yeast thiolase crystals using a polyacrylamide-gel as dialysis membrane, *J. Crsyt. Growth*, **122**, 194–198.

34. Madjid, A.H., Vaala, A.R., Pedulla Jr., J. *et al.* (1972). Diffusion zone process a new method for growing crystals, *Phys. Status Solidi*, **12**, 575–579.

35. Madjid, A.H., Vaala, A.R., Anderson Jr., W.F. *et al.* (1974). Method for growing crystals using a semipermeable membrane, US Patent 3,788,818.

36. Lee, C.Y. Sportiello, M.G., Cape, S.P. *et al.* (1997). Characterization and applications of osmotic dewatering to the crystallization of oligonucleotides, *Biotechnol. Prog.*, **13**, 77–81.

37. Lee, C.Y., McEntyre, S.R. Todd, P. *et al.* (1998). Control of nucleation in oligonucleotide crystallization by the osmotic dewatering method with kinetic water removal rate control, *J. Cryst. Growth*, **187**, 490–498.

38. Kam, Z., Shore, H.B. and Feher, G. (1978). *J. Mol. Biol.*, **123**, 539–555.

39. Cuellar, M.C., Herreilers, S.N. Straathof, A.J.J. *et al.* (2009). Limits of operation for the integration of water removal by membranes and crystallization of L-phenylalanine, *Ind. Eng. Chem. Res.*, **48**, 1566–1573.

40. Smith, J.M., Van Ness, H.C. and Abbott, M.M. (2001). *Introduction to Chemical Engineering Thermodynamics*, McGraw-Hill, New York.

41. Wijmans, J.G. and Baker, R.W.J. (1995). The solution-diffusion model: A review, *J. Memb. Sci.*, **107**, 1–21.

42. Baker, R.W. (2004). *Membrane Technology and Applications*, John Wiley & Sons, Ltd., Chichester, England.

43. Timmer, J.M.K. van der Horst, H.C. and Robbertsen, T. (1993). Transport of lactic acid through reverse osmosis and nanofiltration membranes, *J. Membr. Sci.*, **85**, 205–216.

44. Kapteijn, F., Zhu, W., Moulijn, J.A. *et al.* (2005). *Chemical Industries: Structured Catalysis and Reactors*, CRC, Boca Raton, Vol. 110, pp. 701–748.

45. Rautenbach, R. and Groschl, A. (1993). Fractionation of aqueous organic mixtures by reverse osmosis, *Desalination*, **90**, 93–106.

46. Williams, M.E., Hestekin, J.A., Smothers, C. N. *et al.* (1999). *Ind. Eng. Chem. Res.*, **38**, 3683–3695.
47. Costa, C.A. and Cabral, J.S. (1991). Chromatographic and Membrane Processes in Biotechnology: Institute Proceedings; Springer, New York, pp. 189–190.
48. Bird, R.B., Lightfoot, W.E. and Stewart, E.N. (2002). *Transport Phenomena*, John Wiley & Sons, New York.
49. Welty, J.R., Wicks, C.E., Wilson, R.E. *et al.* (2000). *Fundamentals of Momentum, Heat and Mass Transfer, 4th edn.*, John Wiley & Sons, New York.
50. Kuhn, J., Lakerveld, R., Kramer, H.J.M. *et al.* (2009). Characterization and dynamic optimization of membrane-assisted crystallization of adipic acid, *Ind. Eng. Chem. Res.*, **48**, 5360–5369.
51. Lawson, K.W. and Lloyd, D.R. (1997). Membrane distillation, *J. Membr. Sci.*, **124**, 1–25.
52. Tun, C.M., Fane, A.G. Matheickal, J.T. *et al.* (2005). Membrane distillation crystallization of concentrated salts-flux and crystal formation, *J. Membr. Sci.*, **257**, 144–155.
53. Gryta, M. (2002). Concentration of NaCl solution by membrane distillation integrated with crystallization, *Sep. Sci. Technol.*, **37**, 3535–3558.
54. Curcio, E., Di Profio, G. and Drioli, E. (2003). Recovery of fumaric acid by membrane crystallization in the production of L-malic acid, *Sep. Purif. Technol.*, **33**, 63–73.
55. Di Profio, G., Perrone, G., Curcio, E. *et al.* (2005). Preparation of enzyme crystals with tunable morphology in membrane crystallizers, *Ind. Eng. Chem. Res.*, **44**, 10005–10012.
56. Di Profio, G., Tucci, S., Curcio, E. *et al.* (2007). Controlling polymorphism with membrane-based crystallizers: application to form I and II of paracetamol, *Chem. Mater.*, **19**, 2386–2388.
57. Mulder, M. (1996). *Basic Principles of Membrane Process*, Kluwer Academic, Dordrecht.
58. Di Profio, G., Tucci, S., Curcio, E. *et al.* (2007). Selective glycine polymorph crystallization by using microporous membranes, *Cryst. Growth Des.*, **7**, 526–530.
59. Drioli, E., Curcio E. and Fontananova, E. (2006). 'Mass transfer operation — membrane separations', in Bridgwater, J., Molzahn, M. and Pohorecki, R. (eds.), *Encyclopedia of Life Support Systems (EOLSS)*, Eolss Publishers, Oxford, UK.
60. Di Profio, G., Curcio, E. and Drioli, E. (2008). Progresses on membrane crystallizers, *17th International Symposium on Industrial Crystallization*, Maastricht, The Netherlands, 14–17 Sept. 2008.
61. Lakerveld, R., Kuhn, J., Kramer, H.J.M. *et al.* (2010). Membrane assisted crystallization using reverse osmosis: Influence of solubility characteristics on experimental application and energy saving potential, *Chem. Eng. Sci.*, **65**, 2689–2699.
62. Ji, X., Curcio, E., Al Obaidani, S. *et al.* (2010). Membrane distillation-crystallization of seawater reverse osmosis brines, *Sep. Pur. Techn.*, **71**, 76–82.
63. Curcio, E., Ji, X., Di Profio, G. *et al.* (2010). Membrane Distillation operated at high seawater concentration factors: Role of the membrane on $CaCO_3$ scaling in presence of humic acid, *J. Memb. Sci.*, **346**, 263–269.
64. Banat, F., Jumah, R. and Garaibeh, M. (2002). Exploitation of solar energy collected by solar stills for desalination by membrane distillation, *Renewable Energy*, **25**, 293–305.

65. Koschikowski, J., Wieghaus, M. and Rommel, M. (2003). Solar thermal-driven desalination plants based on membrane distillation, *Desalination*, **156**, 295–304.
66. Ding, Z., Liu, L., El-Bourawi, M.S. *et al.* (2005). Analysis of a solar-powered membrane distillation system, *Desalination*, **172**, 27–40.
67. Mersmann, A. (2001). *Crystallization Technology Handbook, 2nd ed.*, Dekker, New York.
68. Myerson, A.S. (2002). *Handbook of Industrial Crystallization, 2nd ed.*, Butterworth–Heinemann, Boston, MA.
69. Bènet, N., Muhr, H., Plasari, E. *et al.* (2002). New technologies for the precipitation of solid particles with controlled properties, *Powder Technol.*, **128**, 93–98.
70. Charpentier, J.C. (2002). The triplet "molecular processes-product process" engineering: the future of chemical engineering, *Chem. Eng. Sci.*, **57**, 4667–4690.
71. Stankiewicz, A. (2003). Reactive separations for process intensification: an industrial perspective, *Chem. Eng. Process.*, **42**, 137–144.
72. Charcosset, C. and Fessi, H. (2005). Preparation of nanoparticles with a membrane contactor, *J. Membr. Sci.*, **266,** 115–120.
73. Jia, Z., Liu, Z. and He, F. (2003). Synthesis of nano-sized $BaSO_4$ and $CaCO_3$ particles with a membrane reactor: effects of additives on particles, *J. Colloid Interface Sci.*, **266**, 322–327.
74. Chen, G.G., Luo, G.S., Xu, J.H. *et al.* (2004). Membrane dispersion precipitation method to prepare nanoparticles, *Powder Technol.*, **139**, 180–185.
75. Bird, R.B., Stewart, W.E. and Lightfoot, E.N. (1960). *Transport Phenomena*, Wiley, New York, pp. 27–30.
76. Dirksen, J.A. and Ring, T.A. (1991). Fundamentals of crystallization: Kinetic effects on particle size distributions and morphology, *Chem. Eng. Sci.*, **46**, 2389–2427.
77. Jia, Z., Xiao, D., Yang, W. *et al.* (2004). Preparation of perylene nanoparticles with a membrane mixer, *J. Membr. Sci.*, **241**, 387–392.
78. Wu, Q.-S., Sun, D.-M., Liu, H.-J. *et al.* (2004). Abnormal polymorph conversion of calcium carbonate and nano-self-assembly of vaterite by a supported liquid membrane system, *Cryst. Growth Des*, **4**, 717–720.
79. Sun, D.-M., Wu, Q.-S. and Ding, Y.-P. (2006). A novel method for crystal control: synthesis and design of strontium carbonate with different morphologies by supported liquid membrane, *J. Appl. Cryst.*, **39**, 544–549.
80. Sun, D.-M., Zhu, D.-Z. and Wu, Q.-S. (2008). Synthesis and design of $MnCO_3$ crystals with different morphologies by supported liquid membrane, *J. Chem. Crystallogr.*, **38**, 949–952.
81. Martin, C.R. (1996). Membrane-based synthesis of nanomaterials, *Chem. Mater.*, **8**, 1739–1746.
82. Huczko, A. (2000). Template-based synthesis of nanomaterials, *Appl. Phys. A: Mater. Sci. Process*, **70**, 365–376.
83. Hulteen, J.C. and Martin, C.R. (1997). A general template-based method for the preparation of nanomaterials, *J. Mater. Chem.*, **7**, 1075–1087.
84. Schmid, G. (2002). Materials in nanoporous alumina, *J. Mater. Chem.*, **12**, 1231–1238.
85. Charcosset, C., Bernard, S., Fiaty, K. *et al.* (2007). Membrane techniques for the preparation of nanomaterials: Nanotubes, nanowires and nanoparticles — a review, *Dyn. Biochem., Process Biotechnol. Mol. Biol.*, **1**, 15–23.

86. Dobrev, D., Vetter, J. and Neumann, R. (1998). Growth of potassium iodide single-crystals using ion track membranes as templates, *Nucl. Instrum. Methods Phys. Res. B*, **146**, 513–517.
87. Loste, E. and Meldrum, F.C. (2001). Control of calcium carbonate morphology by transformation of an amorphous precursor in a constrained volume, *Chem. Commun.*, 901–902.
88. Loste, E., Park, R.J. Warren, J. *et al.* (2004). Precipitation of calcium carbonate in confinement, *AdV. Funct. Mater.*, **14**, 1211–1220.
89. Meldrum, F.C. and Ludwigs, S. (2007). Template-directed control of crystal morphologies, *Macromol. Biosci.*, **7**, 152–162.
90. Yue, W., Kulak, A.N. and Meldrum, F.C. (2006). Growth of single crystals in structured templates, *J. Mater. Chem.*, **16**, 408–416.

Chapter 2

Theoretical Aspects
in Membrane Crystallization

2.1 Introduction

The concept of membrane crystallization combines the principles of mass and heat transport through porous membranes and the theory of heterogeneous nucleation occurring at the interface with polymeric films. In a membrane crystallizer, the membrane allows the controlled removal of solvent, or the controlled addition of anti-solvent, from/to the crystallizing solution. In addition, morphological parameters such as porosity or roughness, as well as the hydrophobic character of the membrane (expressed in term of contact angle, surface tension, etc.) significantly affect the crystallization kinetics. All these aspects are quantitatively related to each other; the intensity of transmembrane flux affects the rate of supersaturation generation and, consequently, the nucleation rate. Moreover, the Classical Nucleation Theory (CNT) adapted to the case of microporous membranes clearly defines the mathematical relationships existing between the physico-chemical properties of the polymeric surface and the nucleation kinetics.

This chapter aims at providing a detailed formulation of transport and nucleation phenomena at the basis of this innovative technology.

2.2 Mass Transfer

Transport in a typical membrane system is generally described in terms of serial/parallel resistances to the mass transfer from the bulk phase to the membrane interface (on both sides) and across the membrane; this situation is illustrated in Fig. 2.1. An overall coefficient K is conveniently defined as the inverse of the total

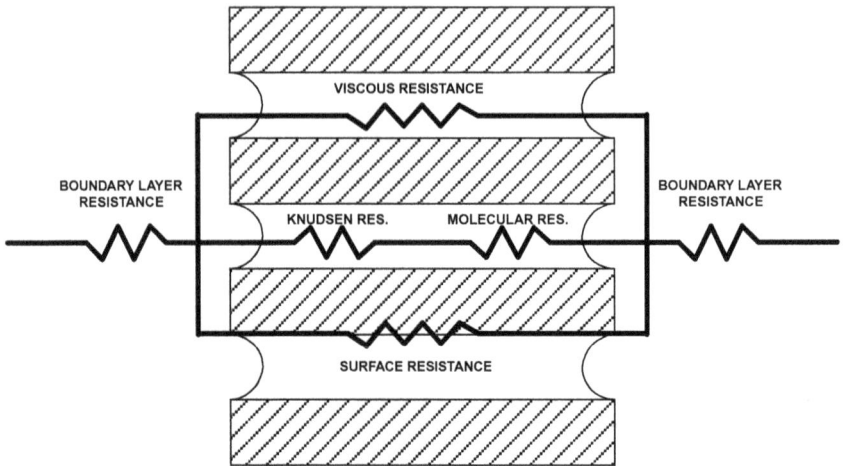

Figure 2.1.   Serial and parallel resistances to mass transport in a membrane crystallization system.

resistance to the mass transport, expressed as a serial combination of the mass transfer coefficient in the feed side (k_f), through the membrane (k_M) and in the distillate/permeate side (k_d):

$$K = \frac{1}{1/k_f + 1/k_M + 1/k_d}. \qquad (2.1)$$

Boundary layers adjacent to the membrane surface generally offer a negligible contribution to the overall mass transfer resistance, whereas diffusion across the polymeric membrane represents the controlling step. Obviously, the resistance to mass transfer on the permeate side can be omitted if the crystallizer is operated with pure condensing fluid in direct contact with the membrane, or if a vacuum is applied. The resistances within the membrane are related to the mechanisms of Knudsen, molecular and surface diffusion, and to viscous transport. A more detailed description of these mechanisms is proposed in Section 2.3.

2.2.1 *Mass transfer across the membrane*

As mentioned before, the mass transfer in a porous medium — assuming surface diffusion negligible[1] — is affected by viscous resistance, Knudsen diffusion resistance (due to collisions between molecules and membrane walls) and/or ordinary diffusion (due to collisions between diffusing molecules).[2] The Knudsen number Kn, defined as the ratio of the mean free path ι of diffusing molecules over the mean pore size of the membrane, discriminates the predominance, coexistence or transition between these different mechanisms. According to the kinetic theory of ideal gases, ι is given by

$$\iota = \frac{k_B T}{P\sqrt{2\pi\sigma^2}}, \tag{2.2}$$

where k_B is the Boltzmann constant ($1.380 \times 10^{-23}\,\mathrm{J\cdot K^{-1}}$), T is the absolute temperature, P the pressure, and σ is the collision diameter of the molecule ($\sigma = 2.7\,\text{Å}$ for water vapor).

For a binary mixture of water vapor in air, the free mean path $\iota_{a/w}$ can be evaluated at the average membrane temperature $\bar{T}$, by[3]

$$\iota_{\frac{a}{w}} = \frac{k_B \bar{T}}{\pi((\sigma_w + \sigma_a)/2)^2 P} \frac{1}{\sqrt{1 + M_w/M_a}}, \tag{2.3}$$

where σ_a (=3.7 Å) and σ_w are the collision diameters, and M_a and M_w are the molecular weight for air and water, respectively. For an average temperature $\bar{T} = 60°C$, Phattaranawik and colleagues calculated a mean free path of 0.11 μm.[4]

In the continuum region, molecule–molecule collisions predominate over molecule-wall collisions and $Kn < 1$. In the Knudsen region (diffusion in nanopores) this situation is reversed: the mean free path of the gas is large with respect to the average membrane pore size, and molecule-wall collisions predominate over molecule-molecule collisions ($Kn > 1$). For applications related to membrane crystallization, ι is comparable to the typical pore size of the microporous membranes used (typically in the range of 0.1 μm to 0.45 μm).

The dusty gas model (DGM) is frequently used for describing the flux of gases through porous media; its most general form (neglecting surface diffusion) is expressed as[5]

$$\frac{J_i^D}{D_{ie}^k} + \sum_{j=i\neq i}^{n} \frac{p_j J_i^D - p_i J_j^D}{D_{ije}^0} = -\frac{1}{RT}\nabla p_i \tag{2.4a}$$

$$J_i^v = -\frac{\varepsilon r^2 p_i}{8RT\tau\mu}\nabla P \tag{2.4b}$$

$$D_{ie}^k = \frac{2\varepsilon r}{3\tau}\sqrt{\frac{8RT}{\pi M_i}} \tag{2.4c}$$

$$D_{ije}^0 = \frac{\varepsilon}{\tau}P D_{ij}^0, \tag{2.4d}$$

where J^D is the diffusive flux of evaporating molecules of the ith species (i.e. the volatile solvent) in mixture with gas j^{th} (usually air), J^v the viscous flux related to a hydrostatic pressure gradient, D^k the Knudsen diffusion coefficient,

D^0 the ordinary diffusion coefficient, p the partial pressure, R the gas constant ($8.314\,\mathrm{J\cdot mol^{-1}\cdot K^{-1}}$), T the absolute temperature, P the total pressure, μ the gas viscosity, r the membrane radius, ε the membrane porosity, and τ the membrane tortuosity. The subscript e indicates the "effective" diffusion coefficient.

Although the DGM is strictly derived for isothermal systems, it is successfully applied whenever a membrane crystallizer is operated under a relatively small thermal gradient; in this case, the average temperature across the membrane is assumed to be the T value.

Despite the low complexity of DGM, and its susceptibility to further reasonable simplification under specific circumstances, the adoption of even simpler empirical correlations is in some cases preferred. The transmembrane flux is thus expressed as a linear function of the vapor pressure difference across the membrane[6]:

$$J = \Theta \Delta p, \qquad (2.5)$$

where Θ is a proportionality coefficient and Δp the vapour pressure gradient at the membrane surfaces. In Eq. (2.5), the coefficient Θ is a function of the membrane morphology (pore size, thickness, porosity, and tortuosity), properties of the vapor transported across the membrane (molecular weight and diffusivity), and operative conditions.

2.2.2 Boundary layer resistances

When solvent molecules of a crystallizing solution are transferred through the membrane, the retained solute accumulates at the membrane interface where its concentration increases. The resulting concentration gradient between membrane surface and bulk of the solution generates a diffusive counterflow that, under steady-state conditions, balances the net convective solute flow through the boundary layer and the flux of solute molecules incorporated in the growing crystals nucleated at the membrane surface. As a consequence, a concentration profile adjacent to the membrane is established; this phenomenon is known as concentration polarization. In a membrane crystallizer, the concentration polarization generally has a limited effect in terms of transmembrane flux, but its influence on nucleation kinetics due to impact on supersaturation value might be relevant.

Referring to Fig. 2.2, and assuming that the non-volatile solute is — at least theoretically — completely retained by the membrane, a mass balance across the feed side boundary gives

$$-D\frac{dc_s}{dx} + J \cdot c_s + d_s G^3 \tau^2 \frac{\rho_c \cdot 1000}{M_w} = 0, \qquad (2.6a)$$

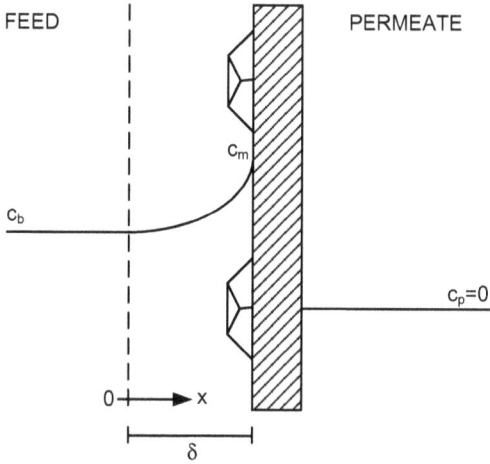

Figure 2.2. Schematic representation of the concentration polarization phenomenon under steady-state conditions.

where D is the diffusion coefficient of solute molecules, c_s the solute concentration, x the spatial coordinate, J the convective volumetric flux through the boundary layer, d_s the surface density of crystals on the membrane (i.e. the number of crystals per membrane area), G the growth rate (G is a function of the actual solute concentration, see details in Section 4.4), τ the residence time within the crystallizer (see Chapter 4), ρ_c the crystal density (i.e. kg/m^3), and M_w the molecular weight of the solute.

Equation (2.6a) is solved under the following boundary conditions:

$$x = 0, c_s = c_b \qquad (2.6b)$$
$$x = \delta, c_s = c_m. \qquad (2.6c)$$

The mass transfer coefficient k_x is defined as

$$k_x = D/\delta. \qquad (2.7)$$

The literature provides several empirical correlations that are useful for determining the mass transfer coefficient, usually expressed in the form[7]

$$Sh = \alpha Re^\beta Sc^\gamma, \qquad (2.8)$$

where

$$Sh = \frac{k_x d_h}{D}$$

Sh: Sherwood number, d_h: hydraulic diameter, D: diffusion coefficient;

$$Re = \frac{\rho v d_h}{\mu}$$

Table 2.1. Predictive equations for mass transfer coefficients in hollow fiber modules.

Equation (lumen-side flow)	Comments	Reference
$Sh = 1.62 \, [d^2 v/(LD)]^{0.33}$	Lévèque equation	8
$Sh = 0.023 \, Re^{0.8} Sc^{0.33}$	Chilton–Colburn, $Re > 10^5$, $Sc > 0.5$	8
$Sh = 0.34 \, Re^{0.75} Sc^{0.33}$	Chilton–Colburn, $10^4 < Re < 10^5$, $Sc > 0.5$	8
$Sh = 0.023 \, Re^{0.875} Sc^{0.25}$	$300 < Sc < 700$	9
$Sh = 0.0149 \, Re^{0.88} Sc^{0.33}$	$Sc > 100$	10
$Sh = 0.023 \, Re^{0.83} Sc^{0.44}$	$0.6 < Sc < 2.5$	11
Equation (shell-side flow)		
$Sh = 1.25 \, (Re \, d_e/L)^{0.93} Sc^{0.33}$	$0.5 < Re < 500$, loosely packed fibers	12
$Sh = 0.019 \, Gz$	$Gz < 60(*)$, closely-packed fibers	13
$Sh = 1.38 \, Re^{0.34} Sc^{0.33}$	$1 < Re < 25$, closely packed fibers	12
$Sh = 0.90 \, Re^{0.40} Sc^{0.33}$	$1 < Re < 25$, loosely packed fibers	12
$Sh = 0.57 \, Re^{0.31} Sc^{0.33}$	$0.01 < Re < 1$, liquid flow perpendicular to the fibers	14

Re: Reynolds number, ρ: fluid density, v: fluid velocity, μ: fluid viscosity;

$$Sc = \frac{\mu}{\rho D}$$

Sc: Schmidt number.

A brief list of mass transfer correlations for newtonian fluids is given in Table 2.1.

$$Gz = \frac{\dot{m}}{\rho \, D \, L}$$

Gz: Graetz number, $\dot{m}$: mass flow rate, L: tube length.

2.2.3 Driving force to mass transfer

In a membrane crystallizer, the driving force to mass transfer is the partial pressure gradient of the evaporating species (typically solvent and/or antisolvent) across the membrane, established by a temperature difference (thermally-driven membrane crystallizer) or by a concentration difference (osmotically-driven membrane crystallizer).

In the case of non-ideal mixtures, the vapor–liquid equilibrium is mathematically described in terms of partial pressure (p_i), vapor pressure of pure i (p_i^0), and

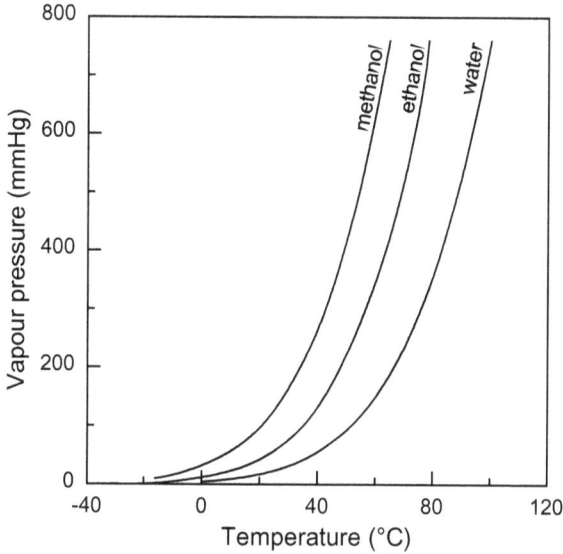

Figure 2.3. The increasing profile of vapor pressure with temperature for different substances.

activity coefficient ξ_i, according to the thermodynamic relationship:

$$p_i = P y_i = p_i^0 a_i = p_i^0 \xi_i x_i, \tag{2.9}$$

where P is the total pressure, a_i the activity of the ith component transported, and x_i and y_i are the liquid and vapor mole fraction, respectively.

The vapor pressure of a pure substance varies with temperature according to the Clausius–Clapeyron equation:

$$\frac{d p^0}{dT} = \frac{p^0 \lambda}{RT^2}, \tag{2.10}$$

where λ is the latent heat of vaporization (for water[15]: 9.7 cal/mol at 100°C), R the gas constant, and T the absolute temperature (Fig. 2.3).

Vapor pressure data (expressed in Pa) are often available in the form of the Antoine equation:

$$p_i^0 = \exp\left[A - \frac{B}{C + T}\right] \tag{2.11}$$

where T is expressed in Kelvin, and A, B, C are constants deduced by regression of experimental measurements (e.g. for water $A = 23.1964$, $B = 3816.44$, $C = -46.13$).[15]

At the pore entrance, the curvature of the vapor–liquid interface is generally assumed to have a negligible effect on the vapor pressure; however, possible influences on the vapor pressure can be evaluated by the Kelvin equation:

$$p^0_{convex\ surface} = p^0 \exp\left[\frac{2\gamma_L}{r\rho RT}\right], \tag{2.12}$$

where r is the curvature radius, γ_L the liquid surface tension, and ρ the liquid molar density.

Activity coefficients can be deduced by a large variety of equations aiming to evaluate the excess Gibbs function of mixtures G^{E}[16,17], this being

$$\ln \xi_i = \left[\frac{\partial \left(nG^E/RT\right)}{\partial n_i}\right]_{T,P,n_j}. \tag{2.13}$$

The Margules equation is the simplest one, and has been found to give similar results to the van Laar equation for several organic solutions. For asymmetric systems showing large positive deviation from ideality, the van Laar equation is preferentially used. The Wilson equation is a powerful tool for systems that do not exhibit liquid–liquid phase splitting.[17]

The non-random two-liquid (NRTL) model is also frequently used in the description of vapor–liquid equilibrium of binary mixtures; for ethanol–water mixtures, Sarti et al. (1993) reported the following expressions[18]:

$$\ln \xi_i = x_k^2 \left[\tau_{ki}\left(\frac{G_{ki}}{x_i + x_k G_{ki}}\right)^2 + \frac{\tau_{ik} G_{ik}}{(x_k + x_i G_{ik})^2}\right] \tag{2.14a}$$

$$\tau_{ik} = a_{ik} + b_{ik}/T; \quad G_{ik} = \exp\left(-\alpha\tau_{ik}\right), \tag{2.14b}$$

where x_i and x_k are the mole fractions of ethanol and water, respectively; the constant values are $a_{ik} = 0.49854$, $b_{ik} = -456.00$, $\alpha = 0.24$.

For diluted aqueous ionic solutions, the value of the activity coefficients can be deduced by the Debye–Hückel theory:

$$\log \xi_{\pm} = -|z_+ z_-| A\sqrt{I}, \tag{2.15a}$$

where $\xi_{\pm}$ is the activity coefficient of the electrolyte, A is a constant which depends on the temperature and solution permittivity, z is the ion valence, and I the ionic strength of the solution, given by

$$I = \frac{1}{2}\sum_i z_i^2 C_i. \tag{2.15b}$$

where C_i is the (molal) concentration of the ith ion; in an aqueous solution at 25°C the constant A is 0.509 $(mol \cdot kg^{-1})^{-1/2}$.[19]

Several attempts have been made to expand the validation of the Debye–Hückel theory to higher concentrations; in this respect, an extended version of the Debye–Hückel equation is given by[20]

$$\ln \xi = -\frac{A^* |z_+ z_-| \sqrt{I}}{1 + a \cdot B \sqrt{I}}, \tag{2.16a}$$

where a is the effective diameter of the ion (expressed in nm) representing the hydrated radius of the ion in the solution (typical values ranges within 3 nm–9 nm), and

$$A^{\phi} = 0.3770 + 4.684 \cdot 10^{-4}(T - 273.15) + 3.74 \cdot 10^{-6}(T - 273.15)^2. \tag{2.16b}$$

B is a constant value equal to 0.3281 at 25°C in water. However, the extended Debye–Hückel equation is only valid to low ionic strength of $I \leq 0.1$ M.

The Pitzer equation improves the activity coefficient (ξ) calculations by including short and long term ionic interaction in calculation[21]; the applicability ionic strength range of Pitzer equation in mixed solutions can be as high as 6 M.

$$
\begin{aligned}
\ln \xi_x = &\left(|z_+|^2 \left\{ -A^{\phi} \left[\frac{\sqrt{I}}{1 + 1.2\sqrt{I}} \right] + \frac{2}{1.2} \ln(1 + 1.2\sqrt{I}) \right.\right. \\
&\left.\left. + \sum_c \sum_a m_c m_a \beta_{ca}'' + \cdots \right\} \right) \\
&\left\{ \cdots + \sum_a \sum_{a''} m_a m_{a''} \phi_{aa''}'' + \sum_c \sum_{c''} m_c m_{c''} \phi_{cc''}'' \right\} \\
&+ \sum_a m_a \left\{ 2\beta_{xa} + \frac{1}{2} \left(\sum_c m_c |z_c| + \sum_a m_a |z_a| \right) C_{xa} \right\} + \cdots \\
&\cdots + \sum_c m_c \left(2\phi_{xc} + \sum_a m_a \psi_{xca} \right) + \sum_a \sum_{a''} m_a m_{a''} \psi_{aa''x}'' \\
&+ |z_+|^2 \sum_c \sum_a m_c m_a C_{ca}
\end{aligned} \tag{2.17a}
$$

$$\ln \xi_y = \left(|z_-|^2 \left\{ -A^\phi \left[\frac{\sqrt{I}}{1 + 1.2\sqrt{I}} \right] + \frac{2}{1.2} \ln \left(1 + 1.2\sqrt{I} \right) \right. \right.$$

$$+ \sum_c \sum_a m_c m_a \beta_{ca}'' + \cdots \left\} \right)$$

$$\left\{ \cdots + \sum_a \sum_{a''} m_a m_{a''} \phi_{aa''}'' + \sum_c \sum_{c''} m_c m_{c''} \phi_{cc''}'' \right\}$$

$$+ \sum_c m_c \left\{ 2\beta_{yc} + \frac{1}{2} \left(\sum_c m_c |z_c| + \sum_a m_a |z_a| \right) C_{yc} \right\} + \cdots$$

$$\cdots + \sum_a m_a \left(2\phi_{ya} + \sum_c m_c \psi_{yca} \right)$$

$$+ \sum_c \sum_{c''} m_c m_{c''} \psi_{cc''y}'' + |z_-|^2 \sum_c \sum_a m_c m_a C_{ca} \qquad (2.17b)$$

For a 1:1 electrolyte (e.g. NaCl):

$$\beta_{ca} = \beta_{ca}^{(0)} + \beta_{ca}^{(1)} \left\{ \frac{2}{\left(2\sqrt{I} \right)^2} \left[1 - \left(1 + 2\sqrt{I} \right) \exp \left(-2\sqrt{I} \right) \right] \right\} \qquad (2.17c)$$

$$\beta_{ca}'' = \frac{\beta_{ca}^{(1)}}{I} \left\{ -\frac{2}{\left(2\sqrt{I} \right)^2} \left[1 - \left(1 + 2\sqrt{I} + \frac{\left(2\sqrt{I} \right)^2}{2} \right) \exp \left(-2\sqrt{I} \right) \right] \right\}$$

$$(2.17d)$$

$$\phi_{ij} = \theta_{ij} + {}^E\theta_{ij} \qquad (2.17e)$$

$$\phi_{ij}'' = {}^E\theta_{ij}'' \qquad (2.17f)$$

$${}^E\theta_{ij} = \left(\frac{|z_+ z_-|}{4I} \right) \left[J \left(x_{ij} \right) - 0.5 J \left(x_{ii} \right) - 0.5 J \left(x_{jj} \right) \right] \qquad (2.17g)$$

$${}^E\theta_{ij}'' = \left(-\frac{{}^E\theta_{ij}}{I} \right) \left(\frac{|z_i z_j|}{8I^2} \right) \left[x_{ij} J''\left(x_{ij} \right) - 0.5 x_{ii} J'' \left(x_{ii} \right) - 0.5 x_{jj} J'' \left(x_{jj} \right) \right]$$

$$(2.17h)$$

$$x_{ij} = 6|z_i z_j| A^\phi \sqrt{I} \qquad (2.17i)$$

$$J(x) = x[4 + 4.581x^{-0.7237} \exp(-0.012x^{0.528})]^{-1} \qquad (2.17j)$$

$$J''(x) = \frac{[4 + 4.581x^{-0.7237} \exp(-0.012x^{0.528})][0.006886x^{0.528} + 1.7237]}{[4 + 4.581x^{-0.7237} \exp(-0.012x^{0.528})]^2}$$

$$(2.17k)$$

$$C_{ca} = \frac{C_{ca}^{\phi}}{2\left(|z_i z_j|\right)^{0.5}}. \qquad (2.17l)$$

In Eq. (2.17), z is the charge of ion, I the ionic strength, m the molality, subscripts x, c, and c'' denote cations, and subscripts y, c, and c'' denote anions. Parameters β are fitted from single-salt data; ϕ account for cation–cation and anion–anion interactions; θ are adjustable parameters defined for each pair of cations and each pair of anions. Superscript E accounts for electrostatic mixing effects of unsymmetrical cation–cation and anion–anion pairs. For water, $A^{\phi}=0.391$ at 25°C and 1 atm total pressure.

Finally, a mean activity coefficient $\xi_{\pm}$ is defined as

$$\xi_{\pm} = \left(\xi_+^{\upsilon_+} \xi_-^{\upsilon_-}\right)^{\frac{1}{\upsilon_+ + \upsilon_-}}, \qquad (2.18)$$

where ξ_+ and ξ_- denote the total activity coefficients of cations and anions, and υ_+ and υ_- the stoichiometric coefficients of the cation and anion in neutral salts.

In an osmotically-driven membrane crystallizer, the solvent is removed from an aqueous solution by contacting the hydrophobic membrane with a hypertonic solution on the opposite side. The vapor diffuses through the membrane under a driving force given by the difference of activity (Fig. 2.4).

Transport in an osmotically-driven membrane crystallizer is not purely a mass transfer operation, because it involves an evaporation at the feed side and a condensation at the stripping side. A temperature difference at the membrane interfaces is thus created, even if the bulk temperatures of the two liquids are equal. Mengual et al. (1993)[22] estimated a temperature difference of about 1°C in aqueous systems concentrated near room temperature by osmotic distillation (OD); although this thermal gradient leads to a negligible decrease of the vapor flux, a possible effect on supersaturation has to be carefully checked.

According to a simplified approach, the transmembrane flux can be expressed in terms of membrane permeability K_m, partial pressures of the solvent (typically water) p'_w at the interface of membrane with feed ($_1$) and stripping ($_2$), and logarithm mean pressure of air entrapped into pores $\bar{p}_{air}$:

$$J = K_m \frac{p'_{w1} - p'_{w2}}{\bar{p}_{air}} \qquad (2.19)$$

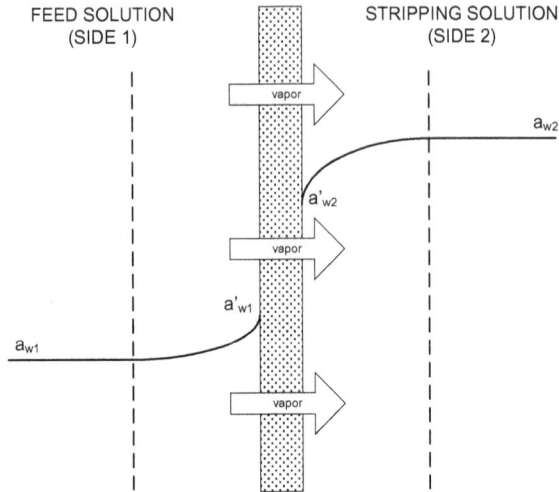

Figure 2.4. Concentration profile in an osmotically-driven membrane crystallizer (a is the activity, subscript w refers to water, apex indicates the value at membrane interface).

The partial pressure of the solvent is related to activity by Eq. (2.9). Referring to a series expansion cut at the first order term, and assuming that the temperature gradient throughout the membrane is small, the following relationship can be derived[23]:

$$J = \frac{K_m}{p_{air}} \left\{ p_w^0 (\bar{T}) \cdot [a'_{w1} - a'_{w2}] - a_w (\bar{T}) \cdot \left[\frac{dp_w^0}{dT} \right]_{\bar{T}} (T'_2 - T'_1) \right\}, \quad (2.20)$$

where p^0 is the vapor pressure of pure liquid, a_w is the water activity, and $\bar{T}$ is the average temperature between the two membrane interfaces.

A common practice is to use a film–theory model to describe the mass transport through boundary layer, and

$$J = k_x \rho \ln \frac{x'}{x} \approx k_x \rho \frac{a_w - a'_w}{\xi_w \bar{x}}, \quad (2.21)$$

where a is the activity, x is the solute mole fraction in the bulk, $\bar{x}$ its logarithmic mean value within the boundary layer, ξ_w the corresponding activity coefficient, k_x the mass transfer coefficient, and ρ the solution density; apex refers to the value at the membrane interface. On the basis of the analogy with OD applications, Table 2.2 reports fewempirical correlations for evaluating k_x.

Temperature and activity gradients can act either in a synergistic or antagonistic way with respect to the promotion of the transmembrane flux. Typical salts chosen as osmotic pressure agents are NaCl (because of its low cost), $MgCl_2$, $CaCl_2$, and

Table 2.2.   Correlations for mass transfer coefficient in OD.

Correlation	α	β	γ	$k_x(10^{-5}$ m/s)	Comment	Reference
$Sh = \alpha Re^{\beta} Sc^{\gamma}$	1.86	0.33	0.33	k_f=0.17; 0.9; 0.37	For NH_3, SO_2, H_2S: $1 \div 20 \times 10^{-3}$ M	24
$Sh = \alpha Re^{\beta} Sc^{\gamma}$	1.62(shell) 1.86(tube)	0.33	0.33	—	Shell: water-sucrose $0 \div 70$wt.% Tube: water-$CaCl_2$: $26 \div 40$wt.%	25

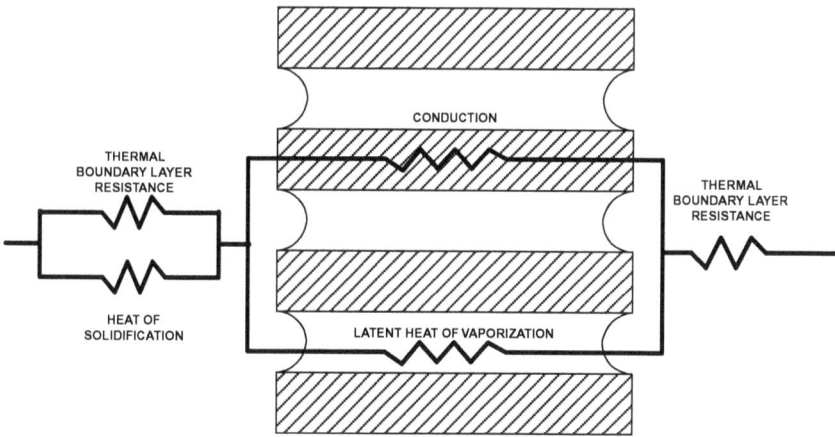

Figure 2.5.   Serial and parallel arrangement of resistances to heat transfer.

$MgSO_4$.[26] In osmotic evaporation carried out at room temperature, transmembrane fluxes generally range between $0.1\,L/m^2{\cdot}h$ and $1\,L/m^2{\cdot}h$.

2.3 Heat Transfer

Simultaneous mass and energy transfer occurring in a membrane crystallizer is usually described in terms of a set of serial and parallel resistances through the boundary layers of the membrane and through the membrane itself. Figure 2.5 illustrates the heat transfer process in the general case; simplifications deriving from the possibility to omit one or more resistances can be made for specific operational configurations.

The presence of boundary layers reduces the energy transport efficiency, and several efforts (use of spacers, turbulence promoters, or turbulent flow), whenever applicable, aim at minimizing such external resistances. With regard to the membrane itself, heat transport occurs in two different ways: conductive heat flow across the polymeric material and latent heat flow related to the transmembrane flux of the vaporized solvent. The total heat flux Q transferred across the membrane is expressed as

$$Q = \left[\frac{1}{h_f + \left(-d_s G^3 \tau^2 \frac{\rho_c \cdot 1000}{M_w} \cdot \Delta H_{sol} \right) / \Delta T_m} \right.$$
$$\left. + \frac{1}{h_m + J \cdot \lambda / \Delta T_m} + \frac{1}{h_p} \right]^{-1} \Delta T, \qquad (2.22)$$

where ΔT is the (bulk) temperature difference among feed and stripping sides, ΔT_m the transmembrane temperature, J the transmembrane molar flux of solvent, λ the molar heat of vaporization, ΔH_{sol} the molar heat of solution, and h_f, h_m and h_p are the heat transfer coefficients on the feed side, membrane, and permeate side, respectively. All other symbols have been previously described in Eq. (2.6a).

2.3.1 Boundary layer resistance and temperature polarization

The most adverse effect due to boundary layer resistances is the creation of a temperature difference between the bulk and the membrane surface where vapor–liquid transition occurs; as a result, the net driving force to transmembrane flux decreases. In addition, the energetic term related to the heat of solidification associated with the formation of a crystal (it is assumed that crystals nucleate at the membrane interface) also influences the temperature and supersaturation profile of the feed boundary layer.

Values of ΔH_f for common salts are reported in Table 2.3.

A temperature polarization coefficient (TPC) is commonly used to quantify the extent of the boundary layer resistances over the total heat transfer resistance. It is

Table 2.3.   Standard enthalpy of solution for common solid compounds.

Solid	ΔH_{sol} (kJ/mol)
NaNO$_3$	35
KCl	17
NaCl	3.9
CaCl$_2$	−82.9
MgSO$_4$	−91.2

Table 2.4. TPCs for various module configurations (data from [6]).

Module and membrane geometry	Fluid-dynamic characteristics	Nusselt number	Hydraulic diameter (mm)	h (W/m^2·K)	TPC
Stirred cell	Re = 8,000	54	50	710	0.2
	Re = 32,000	120	50	1,600	0.4
Parallel plates	Laminar	5.4	8	450	0.1
2 mm thick	Laminar	5.4	2	1,800	5
0.5 mm thick	Laminar	5.4	0.4	8,900	0.4
0.1 mm thick					0.7
Tube	Re = 5,000	29	1	19,000	0.9
	Re = 3,000	20	1	13,000	0.8
	Re = 1,000	4.4	1	2,900	0.54
	Re = 300	4.4	0.3	9,700	0.5
					4
					0.8
Channel	v = 2 m/s	970	500	1,300	0.5
0.5 mm long					1

defined as

$$TPC = \frac{T_f^m - T_p^m}{T_f - T_p},\qquad(2.23)$$

where superscript m indicates the temperature at the membrane surface.

TPC can also be employed to indirectly evaluate the thermal efficiency of the process: it falls between 0.4 and 0.7 for well-designed systems, and approaches unity for mass transfer limited operations. TPC was used by Schofield *et al.* (1987) as a tool in designing membrane distillation systems[6]; Table 2.4 reports calculated TPC for different module configurations and fluid-dynamic conditions.

A large variety of empirical correlations aiming at evaluating boundary layer heat transfer coefficients can be found in the literature. In addition to the previously defined Reynolds number, most of these correlations involve some additional dimensionless numbers:

$$Nu = \frac{hD}{k}$$

Nu: Nusselt number, h: heat transfer coefficient, k: thermal conductivity;

$$Gr = \frac{D^3 \rho^2 g \beta \Delta T}{\mu^2}$$

Gr: Grashof number, g: gravity acceleration, β: thermal expansion coefficient;

$$Pr = \frac{c_p \mu}{k}$$

Table 2.5. Predictive correlations for heat transfer coefficients (tubular conducts).

Equation (tube side)	Comments	Reference
$Nu = 0.13\,Re^{0.64}\,Pr^{0.38}$	Laminar flow	27
$Nu = 0.097\,Re^{0.73}\,Pr^{0.13}$	Laminar flow	27
$Nu = 1.62\left(Re\;Pr\;\left(\dfrac{d}{L}\right)\right)^{0.33}$	Laminar flow, tangential flux	28
$Nu = 0.023\,Re^{0.8}\,Pr^{0.33}\left(\dfrac{\mu}{\mu_w}\right)^{0.14}$	Turbulent flow	28
$Nu = 0.036\,Re^{0.8}\,Pr^{0.33}\left(\dfrac{d}{L}\right)^{0.055}$	Turbulent flow	29
$Nu = 0.027\left(1+\dfrac{6d}{L}\right)Re^{0.8}\,Pr^{0.33}\left(\dfrac{\mu}{\mu_w}\right)^{0.14}$	Turbulent flow	29
$Nu = 1.75\left\{Gz + 0.04\,[(d/L)\;Gr\;Pr]^{0.75}\right\}^{0.33}$	Important influence of free convection	30
$Nu = 0.116\left(Re^{0.66}-125\right)Pr^{0.33}\left[1+(d/L)^{0.66}\right]\left(\dfrac{\mu}{\mu_w}\right)^{0.14}$	Transition region	31
$Nu = 0.298\,Re^{0.646}\,Pr^{0.316}$	Tangential flux	32
$Nu = 0.023\,Re^{0.80}\,Pr^{0.33}$	Vacuum membrane distillation	33
$Nu = 2.0\,Re^{0.483}\,Pr^{0.33}$	Stirred cell	34

Pr: Prandtl number, c_p: specific heat;

$$Gz = \frac{\dot{m}\,c_p}{kL}$$

Gz: Graetz number.

A short review of useful empirical relationships is proposed in Table 2.5.

2.3.2 Heat transfer across the membrane

As mentioned before, the total heat flux Q is transferred across the membrane by two mechanisms: conduction across the membrane material, and as latent heat associated to the vaporized solvent. The balance of energy gives

$$Q = J\lambda(T) - k_m\frac{dT}{dx}, \qquad (2.24)$$

where λ is the enthalpy of vaporization at temperature T, k_m is the thermal conductivity of the membrane ($h_m = k_m/\delta$, where δ is the membrane thickness), and x the longitudinal coordinate. Assuming T_0 as a reference temperature, and considering that temperature inside membrane changes within a few degrees (so that the specific heat can be supposed constant), the vapor enthalpy at a generic temperature T is given by

$$\lambda(T) = \lambda(T_0) + c_{pv}(T - T_0),\qquad(2.25)$$

where c_{pv} is the specific heat of vapour (for water, $c_{pv} = 8.22 + 0.00015T + 0.00000134T^2$ cal/Kmol),[17] and λ the heat of vaporization.

Whereas the latent heat of vaporization is effectively used to promote the permeate flux, the conduction of heat across the membrane is an energetic loss to be minimized. According to previous investigations, the heat loss by conduction represents 20–50% of the overall heat transferred.[35]

In a study on the evaporation efficiency in membrane distillation, Martiez-Diez and colleagues (1999) have carried out experiments on polytetrafluoroethylene (PTFE) microporous membranes and aqueous NaCl solutions[36]; within the range of investigated parameters, the heat lost by conduction per unit transmembrane flux was found to decrease with increasing temperatures and flow rates. The efficiency of MD can also be increased from 8% to 14% by reducing the permeate temperature.[37]

The enthalpy of saturated water vapor and liquid can be evaluated using the following equation in the range of 273K–373K:

$$\lambda(T) = 1.7535T + 2024.3,\qquad(2.26)$$

where T is expressed in K and λ in kJ/kg.

The conduction heat transfer coefficient k_m for a two-phase composite material is generally calculated by the Isostrain model[38]:

$$k_m = (1 - \varepsilon)k_s + \varepsilon k_g,\qquad(2.27)$$

assuming that both polymer (k_s) and gas (k_g) contribute to k_m; ε is the membrane porosity. The following correlations are used to evaluate thermal conductivities of water vapor (k_{H_2O}) and air (k_{air}):

$$k_{H_2O} = 2.72 \cdot 10^{-3} + 5.71 \cdot 10^{-5}T\qquad(2.28a)$$
$$k_{air} = 2.72 \cdot 10^{-3}T + 7.77 \cdot 10^{-5}T.\qquad(2.28b)$$

The values of membrane thermal conductivity found by Eq. (2.27) agree with measured ones within 10%.[39]

The thermal conductivity of polymers strongly depends upon their degree of crystallinity. Data at 296 K for some common hydrophobic polymers span a relatively narrow range: polypropylene (PP): $0.11\,\mathrm{W \cdot m^{-1}K^{-1}}$ to $0.16\,\mathrm{W \cdot m^{-1}K^{-1}}$; polyvinylidenedifluoride (PVDF): $0.17\,\mathrm{W \cdot m^{-1}K^{-1}}$ to $0.19\,\mathrm{W \cdot m^{-1}K^{-1}}$; PTFE:

$0.25\,\mathrm{W\cdot m^{-1}K^{-1}}$ to $0.27\,\mathrm{W\cdot m^{-1}K^{-1}}$.[40] Loss by conduction is reduced by increasing membrane porosity, since the water vapor thermal conductivity is one order of magnitude lower than that of polymeric materials in the range of typical working temperatures.

2.4 Influence of Membrane Morphology on Nucleation Rate

In a membrane crystallizer, the role of the membrane is not merely limited to the promotion and control of the solvent evaporation rate and, ultimately, of the supersaturation.

The following paragraphs show the influence of physico-chemical and morphological parameters such as contact angle, porosity, and roughness on the rate of heterogeneous nucleation. The mathematical description of the system is made according to CNT appropriately revisited.

2.4.1 Effect of contact angle

A crystallizing solution can be imagined as a certain number of solute molecules moving among the molecules of solvent and colliding with each other, so that some of them converge to form clusters (or nuclei). In general, these clusters have a larger probability of being dissolved than of continuing to grow but, under specific conditions, they may achieve a critical size having the same probability (50%) to grow or to dissolve (critical nuclei).

The Gibbs free energy of the crystalline phase, ΔG, is the sum of the contributions from the bulk ΔG_{vol} and from the surface ΔG_{surf}, and the total energy balance gives

$$\Delta G = \Delta G_{vol} + \Delta G_{surf} = -\frac{\Delta\mu}{\Omega}V + \gamma_L A_L - (\gamma_s - \gamma_i)A_{SL}, \qquad (2.29)$$

where $\Delta\mu$ is the chemical potential gradient between the crystalline phase and the mother solution (the driving force to crystallization is a function of supersaturation), Ω the molar volume, and γ_L, γ_i, and γ_s the surface tensions of the nucleus-liquid, nucleus-substrate and liquid-substrate interfaces, respectively. Referring to Fig. 2.6, which shows a sphere cap nucleating on a solid substrate (i.e. a polymeric dense membrane): V is the volume of the sphere cap, A_L is the contact area of the nucleus-solution surface, and A_{SL} is the contact area of the nucleus-membrane surface.

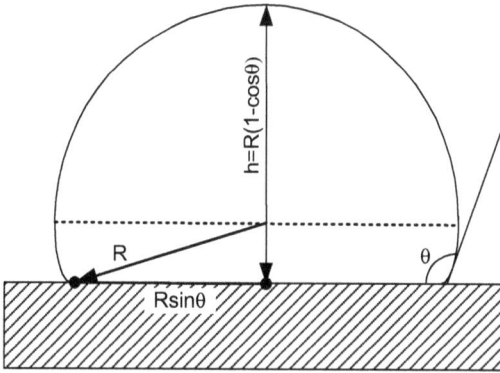

Figure 2.6. Geometry of a sphere cap nucleating on a solid homogeneous surface. The nucleus having radius R forms a contact angle θ with the solid surface.

Simple geometrical considerations give

$$V = \frac{\pi R^3}{3}(1 - \cos\theta)^2(2 + \cos\theta) \tag{2.30a}$$

$$A_L = 2\pi R^2(1 - \cos\theta) \tag{2.30b}$$

$$A_{SL} = \pi R^2 \sin^2\theta. \tag{2.30c}$$

By substituting Eqs. (2.30a–2.30c) into Eq. (2.29),

$$\Delta G = -\frac{\pi R^3}{3}\frac{\Delta\mu}{\Omega}(1 - \cos\theta)^2(2 + \cos\theta)$$
$$+2\pi R^2\gamma_L(1 - \cos\theta) - \pi R^2(\gamma_S - \gamma_i)\sin^2\theta. \tag{2.31}$$

In Eq. (2.31) the gradient of chemical potential $\Delta\mu$ is approximated as

$$\Delta\mu \approx kT\ln\frac{c}{c^*}, \tag{2.32}$$

where k is the Boltzmann constant, T the absolute temperature, c the actual concentration of the mother liquor, and c^* the equilibrium concentration (saturation). The ratio c/c^* defines the supersaturation of the solution.

When nucleation occurs on ideal and non-porous surfaces, the Young equation is suitably used to express the mechanical equilibrium between interfaces:

$$(\gamma_S - \gamma_i) = \gamma_L\cos\theta. \tag{2.33}$$

An exemplificative plot of ΔG as a function of the nucleus radius is shown in Fig. 2.7 at three different values of contact angle. In general, the maximum value for ΔG (indicated as ΔG^*) corresponds to the critical cluster nucleus having radius R^*; further growth of the cluster leads to a decrease in free energy. Clusters of

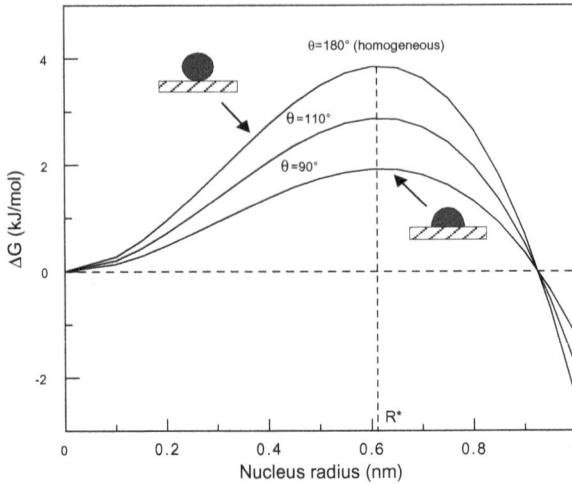

Figure 2.7. Gibbs free energy vs. nucleus radius at three different contact angles ($\theta = 90°$, $110°$, $180°$). The curves reach a maximum in correspondence of the critical radius R^*. Input data: $c/c* = 1.5$; $\Omega = 1.9 \times 10^{-28}$ m^3; γ_L=0.004 J/m^2.

critical size are called critical nuclei; all clusters larger than the size of the critical nucleus will be likely to grow spontaneously. Therefore, crystallization can be considered an activated process, and an energetic barrier ΔG^* must be crossed in order to induce the formation of stable nuclei.[41]

If the maximization condition

$$\frac{d(\Delta G)}{dR} \tag{2.34}$$

is applied to Eq. (2.31), calculations show that, for a critical cluster nucleating in the homogeneous phase (i.e. $\theta = 180°$)

$$R^*_{\text{homogeneous}} = \frac{2\gamma_L\Omega}{\Delta\mu}. \tag{2.35a}$$

By substituting Eq. (2.35a) into Eq. (2.31),

$$\Delta G^*_{\text{homogeneous}} = \frac{16}{3}\pi\gamma_L^3\left(\frac{\Omega}{\Delta\mu}\right)^2, \tag{2.35b}$$

where ΔG^*_{hom} is the value of the Gibbs energy threshold for homogeneous nucleation.

Figure 2.7 demonstrates that the presence of a foreign interface in the crystallizing system — specifically a polymeric membrane with $\theta < 180°$ — decreases the work required to generate a critical nucleus and increases locally the probability of nucleation with respect to other locations in the bulk of the system. This phenomenon is known as heterogeneous nucleation.

In agreement with CNT, calculations show that the Gibbs free energy barrier of a critical nucleus developing on a solid surface is

$$\frac{\Delta G^*_{\text{heterogeneous}}}{\Delta G^*_{\text{homogeneous}}} = \left(\frac{1}{2} - \frac{3}{4}\cos\theta + \frac{1}{4}\cos^3\theta\right). \tag{2.36}$$

From Eq. (2.36), it can be observed that the work of nucleation occurring on a non-porous membrane having a contact angle $\theta = 90°$ is half of that required to form a critical cluster in a homogeneous phase. As expected, the values of the Gibbs free energy barrier for homogeneous and heterogeneous nucleation become equal when $\theta = 180°$.

CNT also elucidates the strict link between ΔG^* and nucleation rate J:

$$J = \Gamma e^{-\frac{\Delta G^*}{kT}}, \tag{2.37}$$

where Γ is a pre-exponential kinetic factor.

Therefore, as a result of the dependence of the energy barrier on the contact angle, the heterogeneous nucleation rate can be controlled by modulating the interaction between the crystallizing solution and the membrane.

When investigating nucleation kinetics, an important parameter to take into account is the induction time, defined as the time elapsed from the attainment of a given supersaturation up to the formation of critical nuclei.[42] As a general rule, a higher nucleation rate results in a shorter induction time. Measurements of induction times confirm the ability of the membrane to promote heterogeneous nucleation. This is particularly evident in the crystallization of proteins, which will be discussed in Chapter 6.

In homogeneous systems, macromolecules display a rather asymmetric and weak bonding configuration at their surfaces, and tend to aggregate in n-mers that diversify the shape and size of lattice units, thus making ordered attachment for growing units less probable.

Crystallization experiments carried out on hen egg white lysozyme (HEWL) demonstrated that, under comparable or even lower supersaturation ratio, induction times for protein crystals grown on PP membranes are lower than those reported in the literature for conventional vapor diffusion techniques.[43]

Membrane surfaces act as a promoter of crystallization by lowering the activation barrier to the nucleation stage, thus allowing molecules to aggregate in conditions of supersaturation that would not be adequate for spontaneous nucleation. The relatively small elapsed time for the appearance of biomolecular crystals demonstrates the existence of the molecules–membrane interactions that favorably affect the mechanisms of nucleation. A short list of biomolecules tested by the authors of this book is reported in Fig. 2.8.

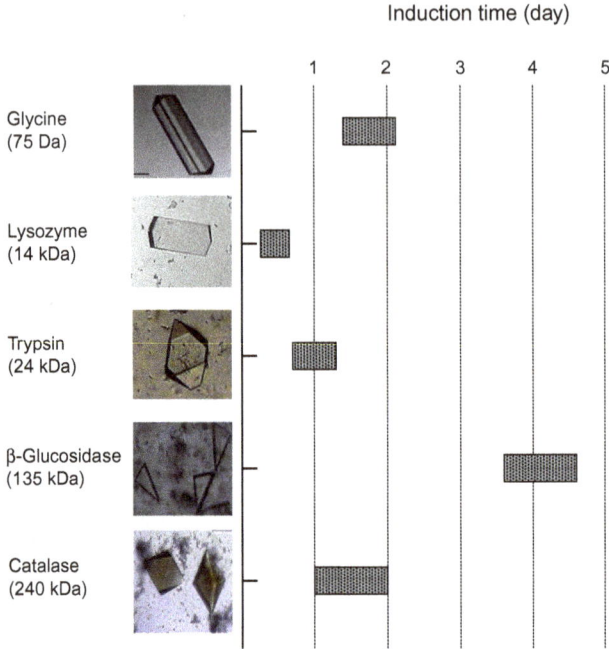

Figure 2.8. Typical ranges of induction times for biomolecules crystallized on microporous hydrophobic polypropylene membranes.

2.4.2 *Effect of membrane porosity*

The influence of membrane porosity on the nucleation kinetics was investigated by Curcio *et al.* (2006).[44] Referring to Eq. (2.29) and considering the geometry of the system under investigation (a sphere cap laying on a porous surface) depicted in Fig. 2.9:

$$V = \frac{\pi R^3}{3} (1 - \cos\theta)^2 (2 + \cos\theta) \tag{2.38a}$$

$$A_L = 2\pi R^2 [(1 - \cos\theta) + \varepsilon (1 + \cos\theta)] \tag{2.38b}$$

$$A_{SL} = \pi R^2 (1 - \varepsilon) \sin^2\theta \tag{2.38c}$$

where R is the radius of the cluster, θ the contact angle, and ε the porosity of the membrane; in Eq. (2.38a), the volume of the small sphere caps penetrating into the pores was considered negligible with respect to the overall volume of the droplet nucleating on the surface due to the hydrophobicity of the membrane.

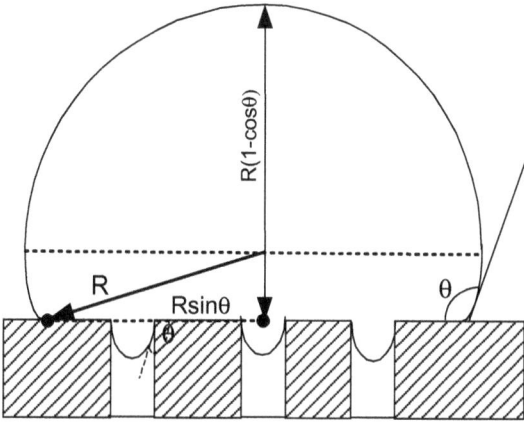

Figure 2.9. Geometry of a sphere cap nucleating on a porous membrane.

If Eqs. (2.38a)–(2.38c) are substituted into Eq. (2.29), then

$$\Delta G = -\frac{\pi R^3}{3}\frac{\Delta\mu}{\Omega}(1-\cos\theta)^2(2+\cos\theta)$$
$$+ 2\pi R^2\gamma_L\left[(1-\cos\theta)+\varepsilon(1+\cos\theta)\right]$$
$$+\cdots-\pi R^2(\gamma_S-\gamma_i)(1-\varepsilon)\sin^2\theta. \tag{2.39}$$

In the case of nucleation on porous membranes, Eq. (2.33) is not applicable because it is strictly valid only for ideal and non-porous surfaces.

For a porous membrane, a modified form of the Young equation correlating the surface porosity to the measured and equilibrium contact angles can be used:

$$(\gamma_S-\gamma_i) = \gamma_L\left(\cos\theta + \frac{4\varepsilon(1+\cos\theta)}{(1-\varepsilon)(1-\cos\theta)}\right). \tag{2.40}$$

The practical validity of Eq. (2.40) was successfully tested on microporous PTFE membranes.[45] After substituting Eq. (2.40) in the last term of Eq. (2.31), and imposing the maximization condition to the variation of the free Gibbs energy as by Eq. (2.34), the value of radius R^* for a critical cluster nucleated on a surface of a porous substrate is

$$R^* = \frac{2\Omega\gamma}{\Delta\mu}\left[1-\varepsilon\frac{(1+\cos\theta)^2}{(1-\cos\theta)^2}\right] = R^*_{hom}\left[1-\varepsilon\frac{(1+\cos\theta)^2}{(1-\cos\theta)^2}\right]. \tag{2.41}$$

For homogeneous nucleation ($\varepsilon=1$ and $\theta=180°$), as well as for a solid surface ($\varepsilon=0$), the value of R^* is equal to the value of the critical radius expressed by Eq. (2.35a) in agreement with the provisions of CNT.

In order to calculate the energetic barrier to heterogeneous nucleation occurring on a porous membrane, Eq. (2.41) is substituted into Eq. (2.39); after appropriate manipulations,

$$\frac{\Delta G^*_{heterogeneous}}{\Delta G^*_{homogeneous}} = \frac{1}{4}(2 + \cos\theta)(1 - \cos\theta)^2 \left[1 - \varepsilon\frac{(1 + \cos\theta)^2}{(1 - \cos\theta)^2}\right]^3. \quad (2.42)$$

If $\varepsilon = 0$, Eq. (2.42) reduces to the mathematical form describing the heterogeneous nucleation on non-porous surfaces as expressed by Eq. (2.36).

The maximum of the Gibbs free energy function associated to the formation of a critical cluster n, which corresponds to the energy barrier that nuclei must overcome to become stable, is plotted in Fig. 2.10 as a function of membrane porosity and observed contact angle. Points are located on the graph according to the contact angle measured for lysozyme solution (40 mg/ml, 2% w/v NaCl) and porosity of PVDF membranes used.

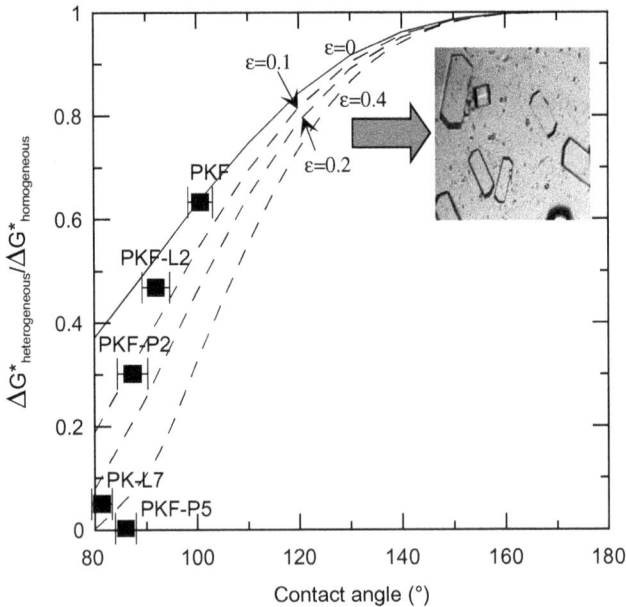

Figure 2.10. Ratio of Gibbs energy barrier for heterogeneous and homogeneous nucleation of HEWL as a function of the membrane porosity and contact angle. Lines are from Eq. (2.42). PKF: poly(vinylidenefluoride-co-hexafluoropropylene); PKF: polyvinylidenefluoride homopolymer; L: LiCl as additive; P: PVP as additive (adapted with permission from.[44] Copyright © 2006 American Chemical Society).

In general, it was observed that the energetic barrier decreases at higher porosity, and therefore nucleation rate increases on highly porous substrates. For instance, the predicted $\Delta G_{heterogeneous}/\Delta G_{homogeneous}$ ratio for PKF-P2 membrane ($\varepsilon = 0.11$) is 0.30, which is 35% lower than the value calculated for an ideal dense polymeric matrix having the same contact angle (87.4°).

For PKF membrane exhibiting low porosity ($\varepsilon = 0.0059$) and high contact angle ($100° \pm 2.5°$), the $\Delta G_{heterogeneous}/\Delta G_{homogeneous}$ ratio differs only by 1.5% with respect to the value of 0.64 predicted by CNT.

Experimental evidence of a favorable effect of porosity on the crystallization rate is well documented in the literature. Luryi and Suhir (1986) observed that the stress energy in forming crystals can be drastically reduced due to pores existing on the substrate surface.[46] The enhancement of macromolecular nucleation on porous silicon surface (PSS) was theoretically investigated by considering the substrate as a fractal object[47]; it was observed that, although the lattice constant of typical protein crystals (in the order of 10 nm or more) was similar to the pore size of the PSS used, pores supported the formation of nuclei for further crystallization on the surface. In the nucleation process associated with capillary condensation of a vapor in a hydrophobic cylindrical pore, it was found that the reduction of the energy barrier predicted by simulation depends on system porosity and pore size.[48]

The nucleation of thaumatin, trypsin, lobster α-crustacyanin, lysozyme, c-phycocyanin, myosin-binding protein-C, and α-actinin actin binding was enhanced in the presence of porous media; non-porous surfaces were less successful at promoting nucleation.[49] Monte Carlo simulations confirm that nucleation out of a filled pore is always faster than on a perfectly smooth surface, and the log of the rate varies almost linearly with pore size.[50]

In principle, the possibility of correlating the membrane morphology to the kinetics of nucleation might offer the opportunity of a more rational design of a crystallization process.

2.4.3 Effect of membrane roughness

In order to adapt the approach provided by CNT to this specific case of interest, let's start again from Eq. (2.29). From simple geometric and trigonometric procedures based on Fig. 2.11:

$$V = \frac{\pi}{3}R^3(1 - \cos\theta)^2(2 + \cos\theta) + \pi R^2 h \sin^2\theta - \frac{\pi}{3}ms^2 h \quad (2.43a)$$

$$A_L = 2\pi R^2(1 - \cos\theta) \quad (2.43b)$$

$$A_{SL} = \pi R^2 \sin^2\theta - m\pi s^2 + m s \sqrt{s^2 + h^2} \quad (2.43c)$$

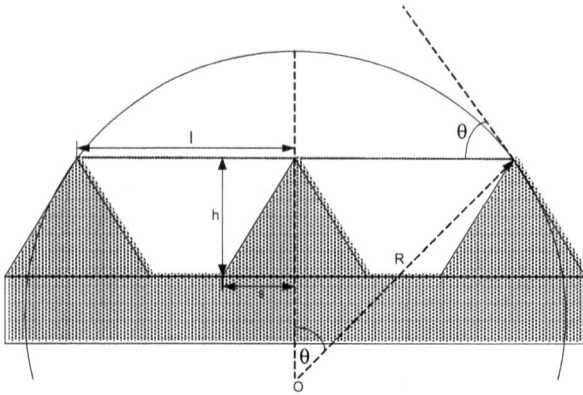

Figure 2.11.   Geometry of a sphere cap nucleus on a rough surface.

where m is the number of asperities on the surface, h is the profile height, l is the distance between two consecutive peaks, s is the half-base length of an asperity, and all other symbols as previously defined.

As stated before, the Young equation is only valid for solid, ideal and smooth surfaces; therefore, the wetting behavior of rough surfaces is described by the Wenzel equation:

$$(\gamma_s - \gamma_i) = \frac{1}{r} \gamma_l \cos \theta, \qquad (2.44a)$$

where r is the Wenzel roughness factor ($r > 1$). In the specific case depicted in Fig. 2.11, assuming that asperities on the membrane surface have a conical shape, the roughness factor r can be evaluated as

$$r = \frac{(\pi\, l^2 - m\pi\, s^2) + m\pi\, s\sqrt{h^2 + s^2}}{\pi\, l^2}. \qquad (2.44b)$$

Mathematical elaborations gives:

$$\frac{\Delta G^*_{heterogeneous}}{\Delta G^*_{homogeneous}} = \frac{1}{4} Y^3 (\cos^3 \theta - 3 \cos \theta + 2) \qquad (2.45a)$$

where

$$Y = \frac{r \cos \theta (1 + \cos \theta) - 2}{\cos \theta (1 + \cos \theta) - 2} \qquad (2.45b)$$

Figure 2.12 illustrates the ratio between the energy barrier for heterogeneous nucleation occurring on rough surfaces and the energy barrier for homogeneous nucleation, as obtained from the maximization condition expressed by Eq. (2.34). The

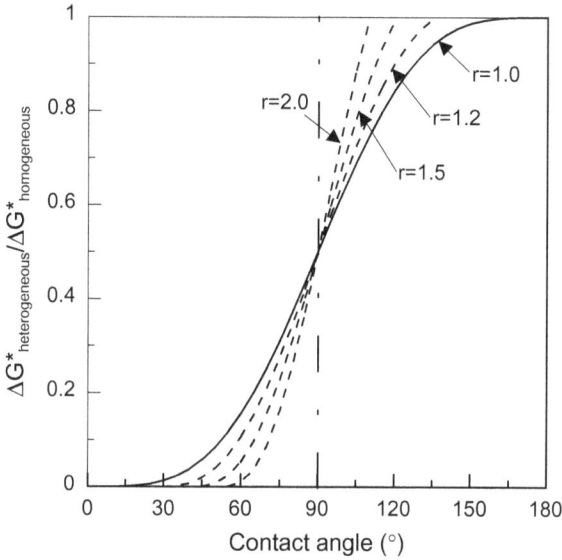

Figure 2.12. Ratio between the activation energy for heterogeneous nucleation process occurring on a rough membrane and the activation energy for homogeneous nucleation, as a function of Wenzel factor and contact angle.

$\Delta G^*_{heterogeneous} / \Delta G^*_{homogeneous}$ increases with the contact angle and, for the limiting value of $\theta = 180°$, the curves converge to 1 (nucleation in homogeneous phase). For a perfectly smooth surface, characterized by a roughness coefficient $r = 1$, the energy ratio is expressed by Eq. (2.42); in particular, for $\theta = 90°$ $\Delta G^*_{heterogeneous} / \Delta G^*_{homogeneous} = 0.5$. If the membrane is hydrophilic, the activation barrier to the formation of critical nuclei decreases at higher surface roughness, thus resulting in an exponentially faster nucleation process; an inverse behavior is observed for hydrophobic membranes.

The literature confirms these theoretical findings. The nucleation rate of calcium sulphate dihydrate was increased by the roughness of heated stainless steel surfaces where $CaSO_4 \cdot 2H_2O$ is deposited.[51] Analogous observations were reported about the influence of surface topography of copper sheet on the heterogeneous nucleation of isotactic polypropylene (iPP) at the iPP/Cu interface.[52] It was also proved that an increase of the wettability and roughness of mica surfaces promotes non-specific and local interactions that contribute to the enhanced nucleation rate of proteins.[53] HEWL crystallization tests carried out on poly(vinylidene fluoride), poly(dimethylsiloxane) and Hyflon membrane surfaces demonstrated that the specific topography significantly affects the nucleation rate.[54]

2.4.4 *Effect of the chemistry of surfaces*

The ability to control the nucleation process occurring on foreign surfaces is essential in many scientific and technological fields. Despite its pervasive incidence, however, the role of the chemistry of surfaces in promoting the heterogeneous nucleation is still not systematically investigated.

Fermani *et al.* (2001) showed that the crystallization of Concanavalin A and Lysozyme on polymeric films containing ionizable groups (such as sulfonated polystyrene, cross-linked gelatin films with adsorbed poly-L-lysine or entrapped poly-L-aspartate and silk fibroin with entrapped poly-L-lysine or poly-L-aspartate) resulted in lower induction times and higher nucleation densities with respect to reference siliconized cover slip.[55]

Apatite nucleation was enhanced by surfaces containing silanol (Si-OH) groups or silicate ions adsorbed on them.[56]

Investigations on non-porous cross-linked poly-4-acryloylmorpholine (PAM), poly-2-carboxyethyl acrylate (PCEA), poly-4-hydroxylbutyl acrylate (PHBA) and polystyrene (PS) substrates used to control the nucleation rate of aspirin showed that polymer films which are polar and contain multiple hydrogen-bonding sites exhibit higher nucleation activity compared to less polar films.[57]

Amorphous surfaces can promote heterogeneous nucleation merely by lowering the surface energy of aggregating molecules (non-specific adsorption), or by structural matching driven by specific polymer–molecule interfacial interactions according to a mechanism analogous to epitaxial growth. The polymer-induced heteronucleation (PIH) capability of polymorph selection, kinetically driven by the preferential aggregation of molecules along specific crystalline facets, confirms these assumptions.[58]

Curcio *et al.* (2014) investigated the mechanisms of acetaminophen (ACM) crystal formation due to specific solute-surface interactions on polydimethylsiloxane (PDMS) with siloxane functional groups (Si-O-Si) forming the backbone of silicones, partially fluorinated elastomer PVDF, PS having weak electron-donor phenyl rings, poly(*n*-buthyl methacrylate) (P*n*BMA) showing ester functionality, polyimide (PI) that includes both hydrogen-bond acceptor imide functionality and carbonyl groups, ethylene/acrylic acid (EAA) copolymer with only the carboxyl moiety.[59]

Measurements of average induction time via probability distribution of crystallization events allowed a preliminary quantitative evaluation of the effect of polymer-solute interactions on the kinetics of heterogeneous nucleation. Assuming that, at constant supersaturation, the formation of a nucleus is an independent and random event, and if v is the average number of nucleation events in a given

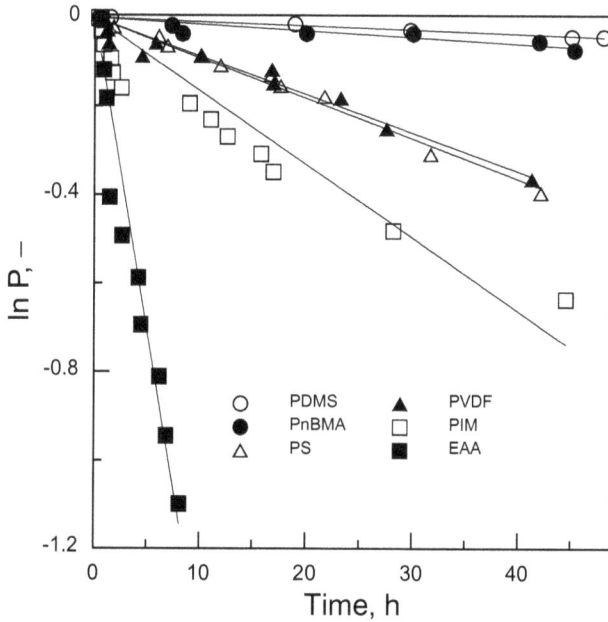

Figure 2.13. ACM crystallization: linear regression of probability data P according to the formula $\ln P = -t/\tau$. Crystallization conditions: $30g_{ACM}/kg\ H_2O$, $10°C$, $800\,\mu L$ solution (reprinted with permission from.[59] Copyright © 2013 American Chemical Society).

interval of time, then the probability P_n of forming n nuclei within the time t is described by the Poisson distribution[60]:

$$P_n = \frac{\nu^n}{n!}e^{-\nu}. \tag{2.46}$$

Finally, the probability of observing no nucleation event (P) within time t is obtained by setting $n = 0$ and $\nu = t/\tau$:

$$P = e^{-t/\tau}. \tag{2.47}$$

From Fig. 2.13, the average induction time (τ) of ACM nucleating in polymer-coated vials at constant supersaturation is calculated by the linear regression of probability data plotting $\ln(P)$ as a function of t; a high value of straight-line slope corresponds to a low induction time.

Powder X-ray diffraction (PXRD) analysis shows the occurrence of preferential orientation (PO) when some of the investigated polymeric surfaces are used as heteronucleants during ACM crystallization. The level of PO is dependent on the strength of the intermolecular interactions occurring at the polymer-crystal interface during nucleation. The strength varies from non-specific adsorption (PDMS,

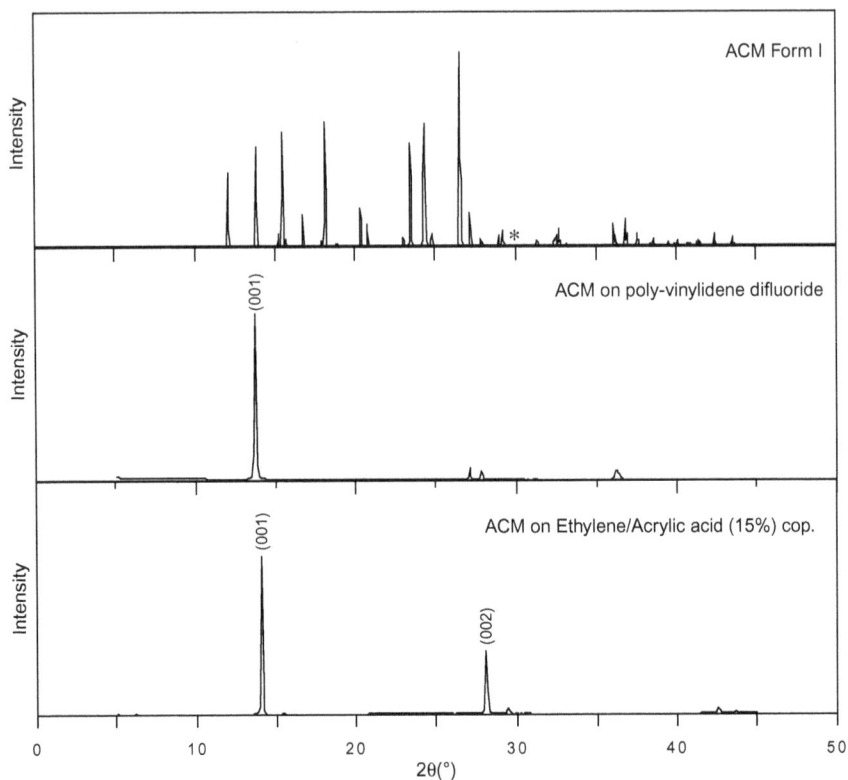

Figure 2.14a. Preferential orientation of ACM crystals nucleated on PVDF (center) and EAA copolymer (bottom) surfaces. Experimental PXRD patterns are compared to the calculated PXRD pattern ACM form I (top). (Reprinted with permission from [59]. Copyright © 2013 American Chemical Society).

PnBMA, and PS) to the oriented arrangement of molecules in the crystalline lattice (EAA, PI, and PVDF). With only one exception, all of the surfaces used in this investigation produced the thermodynamically stable monoclinic, ACM form I. PXRD diffraction patterns for the six investigated polymer surfaces are reported in Figs. 2.14a–2.14c.

PXRD analysis of crystals nucleating on the hydrophobic EAA surface reveals two predominant reflections at 14.1° (001) and 28.1° (002), indicating a strong orientation of ACM crystals along {001}. Referring to Fig. 2.15a, the hydroxyl groups of the ACM molecules aligned along this plane appear perpendicularly oriented towards the polymeric surface, thus indicating that EAA copolymer is interacting through hydrogen bonding.

On the other hand, the PXRD pattern of ACM crystals nucleated on the hydrophobic PDMS surface (contact angle: 95.9° ± 1.6°) exhibits four major

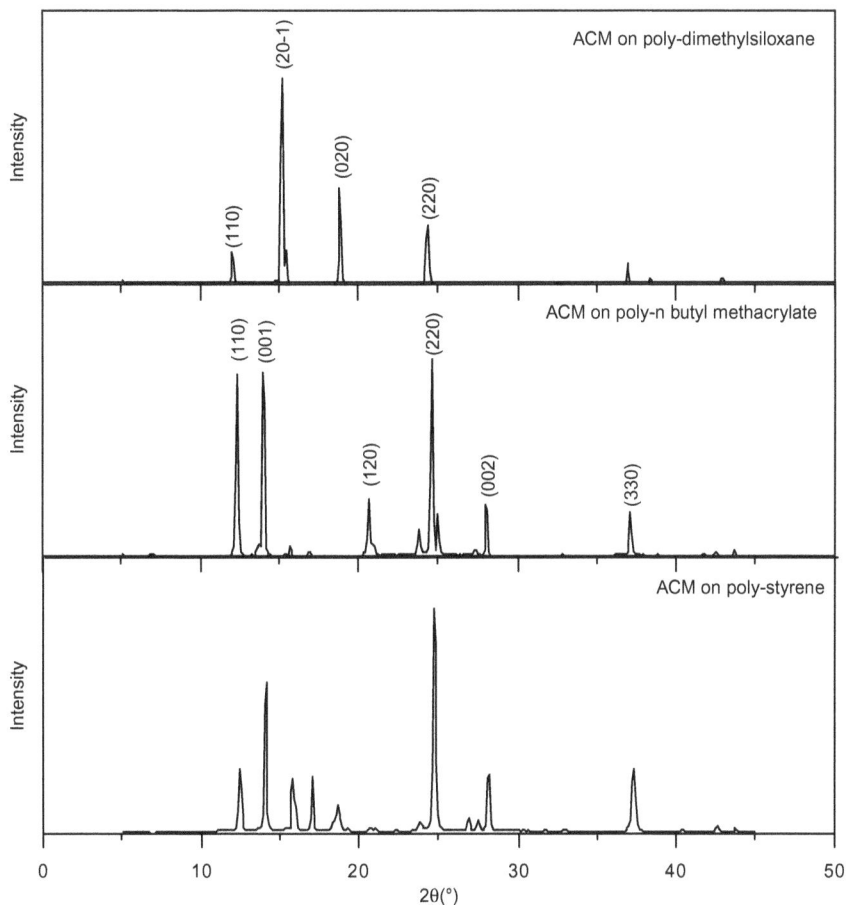

Figure 2.14b. PXRD diffractograms of ACM crystals nucleated on PDMS (top), P*n*BMA (center) and PS (bottom) surfaces showing a low degree of PO along a specific crystallographic plane. (Reprinted with permission from [59]. Copyright © 2013 American Chemical Society).

reflections occurring at 12.1° (110), 15.4° (20-1), 18.9° (020), and 24.4° (220). As a consequence, there is no evidence for a prevalent crystal orientation along a specific crystallographic plane.

Referring to Fig. 2.14b, reflections whose linear combinations correspond to vector direction {110} do not show specific interactions; some ACM molecules are parallel to the polymer surface, others have hydroxyl and methyl groups alternately redirected towards the solid substrate. The main peak at 15.4°, related to orientation along {20-1}, reveals that methyl groups and amide portions in the

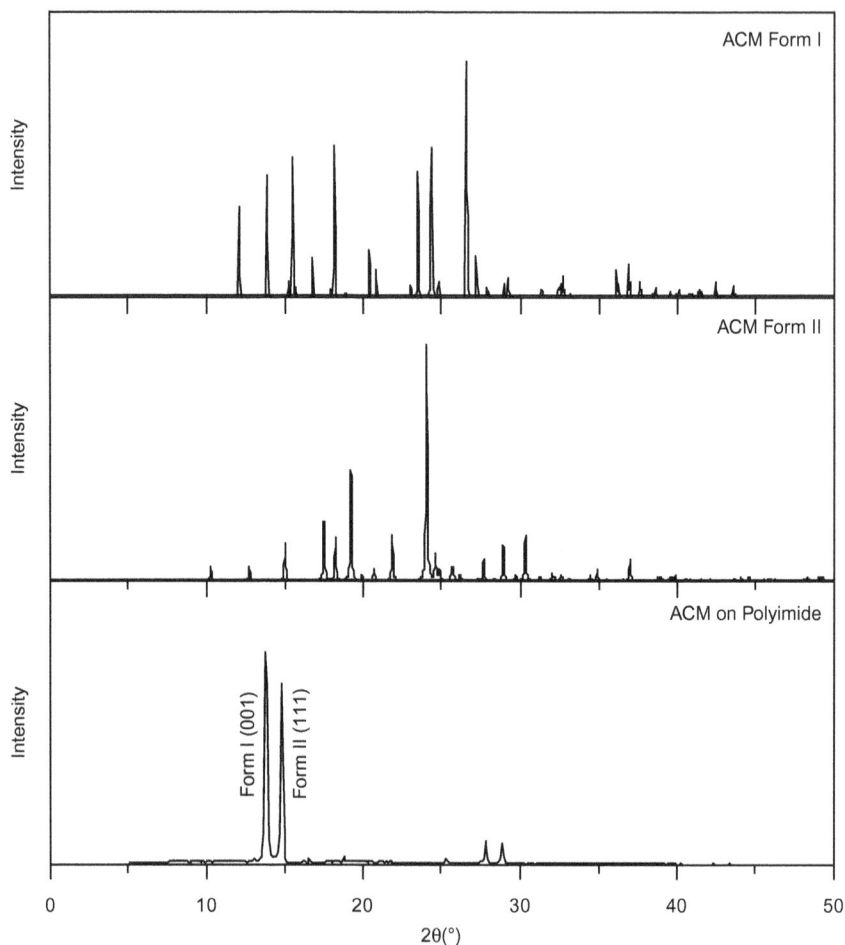

Figure 2.14c. PXRD diffractograms showing concomitant polymorphism in ACM crystals grown on polyimide (PI) surface. The experimental PXRD pattern is compared to the calculated PXRD pattern ACM form I (top) and form II (center). (Reprinted with permission from [59]. Copyright © 2013 American Chemical Society).

ACM molecule are perpendicularly aligned to the PDMS surface (Fig. 2.15c). The slanting orientation of methyl groups with respect to plane (020) are also consistent with the hydrophobic character of PDMS (Fig. 2.15d).

The occurrence of specific polymer–crystal interactions at a molecular level might explain the remarkable difference of the ACM induction time on the EAA surface with respect to the PDMS surface.

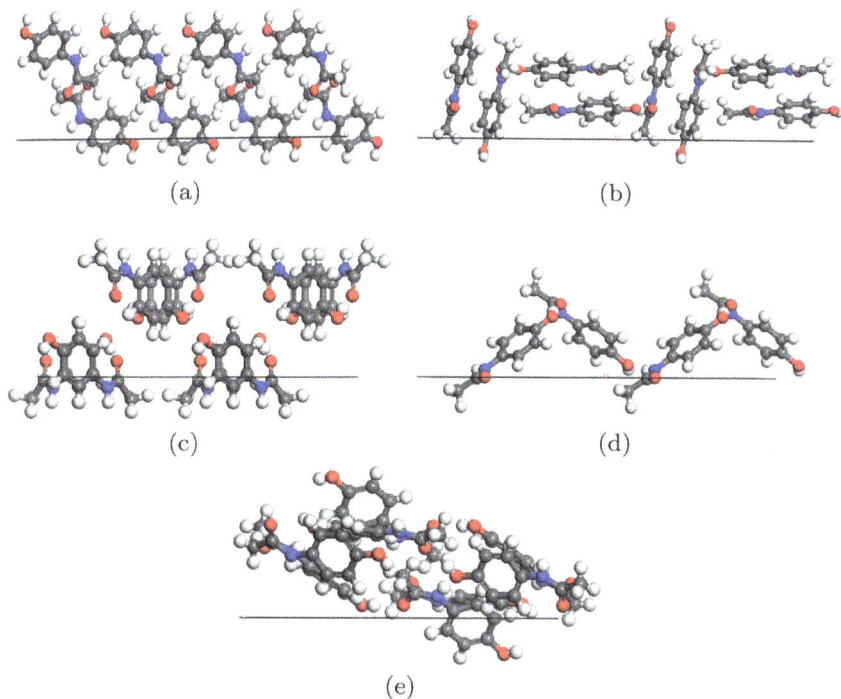

Figure 2.15. View of the molecular arrangement in the crystal structure of (a) monoclinic ACM along (001) face, (b) monoclinic ACM along (110) face, (c) monoclinic ACM along (20 −1) face, (d) monoclinic ACM along (020) face, and (e) orthorhombic ACM along (111) face. ○: Hydrogen, ●: oxygen, ●: nitrogen, ●: carbon. (Reprinted with permission from [59]. Copyright © 2013 American Chemical Society).

The PXRD pattern of ACM crystals grown on PS does not give evidence of PO. The PXRD diffractogram of PnBMA shows a similar behavior, and the presence of significant peaks at 12.1° (110) and 24.4° (220) confirms that mixed hydrophobic-hydrophilic interactions occur along the direction {110} (Fig. 2.15b). The absence of specific hydrogen-bonding interactions is consistent with the observation that the PnBMA surface is dominated by terminal methyl groups of the ester side chain when in contact with water. Crystals nucleated on PVDF surfaces show a surprising orientation along the direction {001} as proved by the predominant peak in the PXRD pattern at 13.9° (001) in 2θ. The strong directional orientation might be attributed to the formation of hydrogen bonds between the electronegative fluorine in the PVDF polymer and the hydrogen in the hydroxyl group of the ACM molecule (Fig. 2.15a).

The PXRD patterns of ACM crystals nucleated on the PI surface revealed the concomitant presence of both the metastable polymorph II (orthorhombic) and the stable form I (monoclinic). In agreement with the hydrophilic character of PI, the major reflection at $13.9°$ is corresponding to the preferential orientation of form I crystals along {001} with hydroxyl groups of ACM oriented towards the polymeric surface (Fig. 2.15a). The major reflection at $15.0°$ is related to the preferential growth of form II crystals along the plane (111) with hydroxyl groups of ACM obliquely oriented towards the substrate (Fig. 2.15e). Interfacial interactions are likely to involve both imide functionality and carbonyl groups in PI.

References

1. Fujii, Y., Kigoshi, S., Iwatani, H. *et al.* (1992). Selectivity and characteristics of direct contact membrane distillation type experiment. I. Permeability and selectivity through dried hydrophobic fine porous membranes, *J. Membrane Sci.*, **72**, 53–72.
2. Kast, W. and Hohenthanner, C.-R. (2000). Mass transfer within the gas-phase of porous media, *Int. J. Heat and Mass Trans.*, **43**, 807–823.
3. Kuhn, H. and Fostering, H. D. (2000). *Principles of Physical Chemistry*, Wiley, New York.
4. Phattaranawik, J., Jiraratananon, R. and Fane, A.G. (2003). Effect of pore size distribution and air flux on mass transport in direct contact membrane distillation, *J. Membrane Sci.*, **215**, 75–85.
5. Mason, E.A. and Malinauskas, A.P. (1983). *Gas Transport in Porous Media: The Dusty-Gas Model*, Elsevier, New York.
6. Schofield, R.W., Fane, A.G. and Fell, C.J.D. (1987). Heat and mass transfer in membrane distillation, *J. Membrane Sci.*, **33**, 299–313.
7. Gekas, V. and Hallstrom, B. (1987). Mass transfer in the membrane concentration polarization layer under turbulent cross flow. I. Critical literature review and adaptation of existing Sherwood correlations to membrane operations, *J. Membrane Sci.*, **30**, 153–170.
8. Bennet, C. and Myers, J. (1982). *Momentum Heat and Mass Transfer*, McGraw-Hill, New York.
9. Deissler, R. (1961). 'Analysis of turbulent heat transfer, mass transfer and friction in smooth tubes and high Prandtl and Schmidt numbers', in Hartnett, J.P. (ed.), *Advances in Heat and Mass Transfer*, McGraw-Hill, New York.
10. Notter, R.H. and Sleicher, C.A. (1971). The eddy diffusivity in the turbulent boundary layer near a wall, *Chem. Eng. Sci.*, **26**, 161–171.
11. Gilliland, E. and Sherwood, T. (1934). Diffusion of vapors into an air stream, *Ind. Eng. Chem.*, **26**, 516–523.
12. Yang, M.C. and Cussler, E.L. (1986). Designing hollow-fiber contactors. *AIChE J.*, **32**, 1910–1915.
13. Wickramasinghe, S.R., Semmens, M.J. and Cussler, E.L. (1992). Mass transfer in various hollow fiber geometries, *J. Membrane Sci.*, **69**, 235–250.

14. Wang, K.L. and Cussler, E.L. (1993). Baffled membrane modules made with hollow fiber fabric, *J. Membrane Sci.*, **85**, 265–278.
15. Reid, R.C., Prausnitz, J.M. and Sherwood, T.K. (1977). *The Properties of Gases and Liquids, 3rd edn.*, McGraw-Hill, New York.
16. Drioli, E., Criscuoli, A. and Curcio, E. (2005). *Membrane Contactors. Fundamentals, Potentialities and Applications*, Elsevier, The Netherlands.
17. Perry, R.H. (1984). *Perry's Chemical Engineering Handbook, 6th edn.*, McGraw-Hill Book Co., Singapore.
18. Sarti, G.C., Gostoli, C. and Bandini, S. Extraction of organic components from aqueous streams by vacuum membrane distillation, *J. Membrane Sci.*, **80**, 21–33.
19. Butler, J. (1982). *Carbon Dioxide equilibria and Their Applications*, Addison-Wesley, California.
20. Sohnel, O. and Garside, J. (1992). *Precipitation Basis Principals and Industrial Applications*, Butterworth–Heinemann, Oxford.
21. Pitzer, K. (1973). Thermodynamics of electrolytes. I. Theoretical basis and general equations, *J. Phys. Chem.*, **77**, 268–276.
22. Mengual, J.I., Ortiz de Zarate, J., Pena, L. *et al.* (1993). Osmotic distillation through porous hydrophobic membranes, *J. Membrane Sci.*, **82**, 129–140.
23. Celere M. and Gostoli, C. (2002). The heat and mass transfer phenomena in osmotic membrane distillation, *Desal.*, **147**, 133–138.
24. Zhang, Q. and Cussler, E.L. (1985). Hollow fiber gas membranes, *AIChE J.*, **31 /9**, 1548–1553.
25. Hogan, P.A. (1996). *Thermal and Isothermal Membrane Distillation*, PhD Thesis, University of New South Wales.
26. Sheng, J. (1993). Osmotic distillation technology and its applications, *Austral. Chem. Eng. Conf.*, **3**, 429–432.
27. Gryta, M., Tomaszewska, M. and Morawski, A.W. (1997). Membrane distillation with laminar flow, *Sep. Purif. Technol.*, **11**, 93–101.
28. Rimura S. and Nakao, S. (1987). Transport phenomena in membrane distillation, *J. Membrane Sci.*, **33**, 285–298.
29. Ozisik, M.N. (1985). *Heat Transfer*, McGraw-Hill, New York.
30. Gryta, M. and Tomaszewska, M. Heat transport in the membrane distillation process, *J. Membrane Sci.*, **144**, 211–222.
31. Mengual, J.I., Khayet, M. and Godino, M.P. (2004). Heat and mass transfer in vacuum membrane distillation, *Int. J. Heat and Mass Trans.*, **47**, 865–875.
32. Tomaszewska, M., Gryta, M. and Morawski, A.W. (1995). Study on the concentration of acids by membrane distillation, *J. Membrane Sci.*, **102**, 113–122.
33. Lawson, K.W. and Lloyd, D.R. (1996). Membrane distillation. I. Module design and performance evaluation using vacuum membrane distillation, *J. Membrane Sci.*, **120**, 111–121.
34. Sudoh, M., Takuwa, K., Iizuka H. *et al.* (1997). Effects of thermal and concentration boundary layers on vapor permeation in membrane distillation of aqueous lithium bromide solution, *J. Membrane Sci.*, **131**, 1–7.
35. Fane, A.G., Schofield, R.W. and Fell, C.J.D. (1987). The efficient use of energy in membrane distillation, *Desal.*, **64**, 231–243.

36. Martinez-Diez, L., Florido-Diaz, F.J. and Vasquez-Gonzalez, M.I. (1999). Study of evaporation efficiency in membrane distillation, *Desal.*, **126**, 193–198.
37. Calabrò, V., Drioli, E. and Matera, F. Membrane distillation in the textile wastewater treatment, *Desal.*, **83**, 209–224.
38. Warner, S.B. (1995). *Fiber Science*, Prentice-Hall, Englewood Cliffs, NJ.
39. Schofield, R.W., Fane, A.G. and Fell, C.J.D. (1990). Gas and vapour transport through microporous membranes. II: Membrane distillation, *J. Membrane Sci.*, **53**, 173–185.
40. Brandrup, J. and Immergut, E.H. (1989). *Polymer Handbook, 3rd edn.*, Wiley, New York.
41. Volmer, M. and Weber, A. (1926). Keimbildung in ubersettigten gebilden, *Z. Phys. Chem.*, **119**, 277–301.
42. Garside, J., Mersmann, A. and Nyvlt, J. (2002). *Measurement of Crystal Growth and Nucleation Rates*, IChemE, Rugby, UK.
43. Di Profio, G., Curcio, E., Cassetta, A. *et al.* (2003). Membrane crystallization of lysozyme: kinetic aspects, *J. Cryst Growth*, **257**, 359–369.
44. Curcio, E., Fontananova, E., Di Profio, G. *et al.* (2006). Influence of the structural properties of poly(vinylidene fluoride) membranes on the heterogeneous nucleation rate of protein crystals, *J. Phys. Chem. B*, **110**, 12438–12445.
45. Troger, J., Lunkwitz, K. and Burger, W. (1997). Determination of the surface tension of microporous membranes using contact angle measurements, *J. Colloid and Interf. Sci.*, **194**, 281–286.
46. Luryi, S. and Suhir, E. (1986). New approach to the high quality epitaxial growth of lattice-mismatched materials, *Appl. Phys. Lett.*, **49**, 140–142.
47. Stolyarova, S., Baskin, E., Chayen, N.E. *et al.* (2005). Possible model of protein nucleation and crystallization on porous silicon, *Phys. Status Solidi A*, **8**, 1462–1466.
48. Saugey, A., Bocquet, L. and Barrat, J.L. (2005). Nucleation in hydrophobic cylindrical pores: A lattice model, *J. Phys. Chem. B*, **109**, 6520–6526.
49. Chayen, N.E., Saridakis, E. and Sear, R.P. (2006). Experiment and theory for heterogeneous nucleation of protein crystals in a porous medium, *P. Natl. Acad. Sci. USA*, **103**, 597–601.
50. Page, A.J. and Sear, R.P. (2006). Heterogeneous nucleation in and out of pores, *Phys. Rev. Lett*, **97**, 065701/1–4.
51. Gunn, D.J. (1980). Effect of surface roughness on the nucleation and growth of calcium sulphate on metal surfaces, *J. Cryst. Growth*, **50**, 533–537.
52. Lin, C.W., Ding, S.Y. and Hwang, Y.W. (2001). Interfacial crystallization of isotactic polypropylene molded against the copper surface with various surface roughnesses prepared by an electrochemical process, *J. Mater. Sci.*, **36**, 4943–4948.
53. Falini, G., Fermani, S., Conforti, G. *et al.* (2002). Protein crystallisation on chemically modified mica surfaces, *Acta Crystallogr. D*, **58**, 1649–1652.
54. Curcio, E., Curcio, V., Di Profio, G. *et al.* (2010). Energetics of protein nucleation on rough polymeric surfaces, *J. Phys. Chem. B*, **114**, 13650–13655.
55. Fermani, S., Falini, G., Minnucci, M. *et al.* (2001). Protein crystallization on polymeric film surfaces, *Cryst. Growth*, **224**, 327–334
56. Miyaji, F., Iwai, M., Kokubo, T., *et al.* (1998). Chemical surface treatment of silicon for inducing its bioactivity, *J. Mater. Sci. — Mater. Med.*, **9**, 61–65.

57. Diao, Y., Myerson, A.S., Hatton, T.A. *et al.* (2011). Surface design for controlled crystallization: The role of surface chemistry and nanoscale pores in heterogeneous nucleation, *Langmuir*, **27**, 5324–5334.
58. López-Mejías, V., Kampf, J.W. and Matzger, A.J. (2012). Nonamorphism in flufenamic acid and a new record for a polymorphic compound with solved structures, *J. Am. Chem. Soc.*, **134/24**, 9872–9875.
59. Curcio, E., López-Mejías, V., Di Profio, G. *et al.* (2014). Regulating nucleation kinetics through molecular interactions at the polymer–solute interface, *Cryst. Growth Des.*, **14/2**, 678–686.
60. Alhalaweh, A., Vilinska, A., Gavini, E. *et al.* (2011). Surface thermodynamics of mucoadhesive dry powder formulation of zolmitriptan, *AAPS PharmSciTech.*, **12/4**, 1186–1192.

Chapter 3

Membrane Materials

3.1 Introduction

The membrane itself represents the core of a membrane crystallizer. Although a quite large variety of membranes exists (depending on their different morphology, topology, mass and energy transport properties, selectivity, and separation mechanism), microporous hydrophobic membranes are usually required for membrane crystallization (MCr) operations. However, in principle, any surface is suitable for promoting a controlled heterogeneous nucleation phenomenon. According to the most traditional classification, synthetic membranes are divided into organic (polymeric) and inorganic membranes; within the same class, different properties are originated by dissimilar raw materials or preparation methods.

At present, due to their easy processability on industrial scale and possibility to properly modulate their physico-chemical properties, polymeric membranes have attracted much more interest than inorganic ones. The most significant exception probably concerns the use of ceramic membranes in high-temperature separation and emulsification processes.

3.2 Membrane Polymers

The selection of a polymeric material is preliminarily driven by the necessity to achieve a high chemical and thermal stability. Moreover, preparation methods and related characteristics of the resulting membranes critically limit the choice of materials. It is beyond the scope of this book to give details on such an extremely complex matter, and readers are referred to specific handbooks and literature papers in this field.

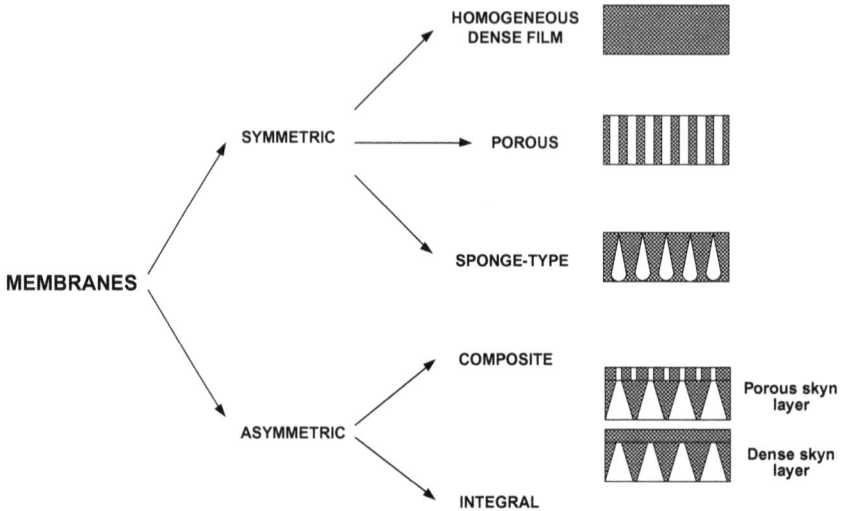

Figure 3.1. Classification of artificial membranes according to their physical structure.

The morphology of a membrane is a critical subject, and it is largely determined by the preparation procedures: in general, the first classification level discriminates symmetric and asymmetric membranes (Fig. 3.1).

Symmetric membranes, characterized by almost uniform properties, include dense homogeneous films or structures having straight or sponge-like pores; such microporous membranes are widely employed in membrane contactor technology (membrane distillation (MD), MCr, membrane absorption, stripping and extraction, support for liquid membranes, membrane emulsification). Asymmetric membranes exhibit a thin dense or porous skin layer (typically less than 1 μm) that provides the desired selectivity on the top of a highly porous sub-layer (having a typical thickness of 100 μm–200 μm); the porous substructures supply the necessary mechanical properties and offer the advantage of a low resistance to the mass transport through the membrane. In integral asymmetric membranes, both the skin layer and the porous support are manufactured from the same material; the opposite situation characterizes the composite asymmetric membranes.

Although most efforts during the past several decades have been devoted to the development of hydrophilic membranes driven by the industrial interest on traditional filtration processes, nowadays more attention is paid to emerging processes such as MD, which requires hydrophobic membranes.[1] Among hydrophobic materials, fluoropolymers constitute a unique class of materials with a combination of interesting properties that has attracted significant attention over the past few decades.[2,3] Generally, these polymers have high thermal stability, improved chemical resistance, and low surface tension because of the low polarizability, strong electronegativity of the fluorine atoms, small van der Waals radius (1.32 Å) and strong C-F bond (485 kJ·mol^{-1}). Due to these outstanding properties, fluoropolymers are an excellent choice for various applications, including MCr. Recently, many fluorinated polymers and copolymers have been widely employed to fabricate membranes, especially with the development of the thermally-induced phase separation (TIPS) method, which can be applied to polymers that are not soluble in most common solvents at room temperature. The important fluoropolymers for membrane operations are listed in Table 3.1 along with their chemical structures.

Table 3.1. Typical materials for microporous hydrophobic membranes.

Polymer	Chemical structure	Main characteristics
Polypropylene (PP)		Chemically resistant; hydrophobic
Poly(vinylidenefluoride) (PVDF)		Highly temperature-resistant; inherently hydrophobic; good processability
Poly(tetrafluoroethylene) (PTFE)		Highly temperature- and chemical- (acid)-resistant; cannot be irradiated; inherently hydrophobic
Poly(ethylene chlorotri-fluoroethylene) (PECTFE)		High basis and solvent resistance behavior
HYFLON		Highly hydrophobic, temperature-resistant, good processability
TEFLON		Highly hydrophobic, temperature-resistant, good processability
CYTOP		Highly hydrophobic, temperature-resistant, good processability

3.2.1 Poly(vinylidene fluoride)

Homopolymers of PVDF are semi-crystalline and long chain macromolecules which contain 59.4 wt.% fluorine and 3 wt.% hydrogen.[4,5] They typically have a crystallinity of 35%–70%, depending on the preparation and thermal mechanical history. Molecular weight, molecular weight distribution, extent of irregularities along the polymer chain, crystallinity and polymorphism, are the major factors influencing the properties of PVDF. The high level of crystallinity yields PVDF stiffness, toughness, and creep-resistance. The crystalline phase of PVDF has three different molecular conformations and five distinct crystal polymorphs. The crystal polymorphs are: α (phase II), β (phase I), γ (phase III), δ, and ε, among which the α-, β-, and γ-phases are the most frequent.[5] The α-form is kinetically favorable, while the β-form is the most thermodynamically stable form.

Because of the combination of good properties and processability, PVDF has been widely used in the fabrication of hydrophobic membranes with high performance. PVDF is inert to various solvents, oils and acids, and shows low permeability to gases and liquids. PVDF is available in a wide range of melt viscosities as powders and pellets to fulfill typical fabrication requirements. All common extrusion and molding techniques can be used to process it into different shapes. The main preparation methods for microporous PVDF membranes are non-solvent induced phase separation (NIPS) and TIPS. In NIPS process, PVDF is dissolved in a proper solvent close to room temperature, and a non-solvent is employed to induce phase separation. For the TIPS process, PVDF is melt blended with diluents at a high temperature, and phase separation is induced by decreasing the temperature. For both methods, the selection of suitable solvents/diluents has a crucial effect on the morphology, structures, properties and performances of the resulting membranes. The NIPS method is easier to realize if the polymer is soluble in a suitable solvent, while the TIPS method can fabricate fluoropolymer membranes with higher mechanical strength and narrower pore-size distribution.

An additional recent method for fabricating PVDF membranes is vapor-induced phase separation (VIPS). In this process, a polymer solution is cast into a film or spun into a desired shape and exposed to a vapor of a non-solvent. Phase separation of the polymer solution is induced by the penetration of non-solvent vapor into the solution.[6,7] This method is more efficient in creating highly porous PVDF membranes with pore sizes in the micron and sub-micron range.

Moreover, a simple and effective freezing method has been explored to fabricate PVDF microporous membranes with high water fluxes and excellent mechanical properties.[8,9] The polymer solution was cast onto a glass plate, and the solidification of casting solutions was achieved by a freezing method. The frozen solvent was exchanged, and PVDF microporous membranes were formed. This method can yield PVDF membranes with some of the characteristics typically induced by the TIPS method while avoiding high temperatures. The mechanical properties of PVDF membranes were drastically enhanced by this freezing method.

Recently, electrospinning process is receiving increasing attention for preparing PVDF nanofiber membranes,[10–12] having contact angle higher ($\sim$130°C) than that of conventional phase inversion membranes ($\sim$80°C). PVDF nanofiber membrane were tested in desalination applications by air-gap membrane distillation (AGMD), exhibiting high permeate flux and NaCl rejection, and long-term stability. In addition, coloumbic forces stimulated by electrospinning may promote the formation of β-phase, which is beneficial for use in ionic membranes.[13] Applicable solvents for NIPS[14] are N,N-dimethyl acetamide (DMAc), N,N-dimethyl formamide (DMF), dimethyl sulfoxide (DMSO), hexamethyl phosphoramide (HMPA), N-methyl-2-pyrrolidone (NMP), tetramethylurea (TMU), triethyl phosphate (TEP), trimethyl phosphate (TMP), acetone (Ac), and tetrahydrofuran (THF); applicable diluents for TIPS[15] are dimethyl phthalate (DMP), diethyl phthalate (DEP), dibutyl phthalate (DBP), dihexyl phthalate (DHP), glycerin triacetate (GTA), ethyl acetoacetate (EAA), propylene glycol carbonate (PGC), diphenyl ketone (DPK), diphenyl carbonate (DPC), cyclohexanone, camphor, and butyrolactone.

Although PVDF has been widely employed in membrane processes, many copolymers of PVDF have been developed to obtain specific properties suitable for new emerging membrane processes. Properties such as crystallinity, melting point, glass transition temperature, stability, elasticity, permeability, and chemical reactivity may be changed as a result of copolymerization.[16] Poly(vinylidene fluoride-co-tetrafluoroethylene) (P(VDF-co-TFE)), poly(vinylidene fluoride-co-hexafluoropropene) (P(VDF-co-HFP)), poly(vinylidenefluoride-co-chlorotri fluoroethylene) (P(VDF-co-CTFE)), poly(vinylidene fluoride)-graft-poly-(styrenesulfonic acid) (PVDF-g-PSSA), and poly(vinylidene fluoride–tri fluoroethylene) (P(VDF-TrFE)) are important PVDF copolymers in membrane manufacturing.

3.2.2 Poly(tetrafluoroethylene)

PTFE constitutes 60 wt.% of the world fluoropolymer consumption.[17] Compared to PP and PVDF, PTFE has the lowest surface energy, which determines the highest hydrophobicity. Furthermore, the exceptional combination of thermal resistance, chemical stability and low surface tension makes PTFE an excellent material for MD and MCr . PTFE macromolecules are long and unbranched, so that polymer typically exhibit a high degree of crystallinity (from 92% to 98%). CF_2 groups form the polymer chain, and the whole chain twists into a helix as in a length of rope; the carbon chain twists because the large fluorine atoms prohibit a planar zigzag conformation. The fluorine atoms surrounding the carbon backbone act as a protective sheath, so determining a high melting point, high thermal stability, very good chemical resistance, low coefficient of friction, and low dielectric constant. Main disadvantages of PTFE are that the material does not flow easily even above its melting point 327°C,[18] and that it is not possible to dissolve PTFE in any substance at room temperature. The high molecular weight of standard PTFE, which is in the range of 1×10^6 to 5×10^6, is the main reason for the extremely high melt viscosity.

PTFE is very hydrophobic, which is beneficial for membrane processes in which vapor or gas passes through the membrane. PTFE membranes have been adopted and investigated in MD and pervaporation (PV).[19] Among the commercially available membranes, PTFE has a sufficiently high contact angle to avoid wettability or degradation when in contact with the amine solvents in membrane gas absorption (MGA) process.[20] PTFE is also employed to sputter other polymers in order to enhance their hydrophobicity.[21] Despite these advantages, PTFE membranes are quite expensive. Fabrication of PTFE products is difficult since its poor processing properties prohibit conventional phase inversion or melt spinning methods.[22] At present, commercial PTFE membranes are usually produced by extrusion, rolling and stretching, or sintering procedures.

Recently, a spinning solution was prepared by blending poly(vinyl alcohol) (PVA) with a PTFE.[23] The solution was electrospun into composite nanofibers, then sintered to obtain a PTFE porous membranes . The membranes exhibited good thermal stability and strong hydrophobicity.[24] Unique 'worm-like' structures (crystals of PTFE) were observed at different PTFE/PVA mass ratio, and the growth of these crystals improved both hydrophobicity and mechanical strength of the PTFE membrane.[25] In addition, a novel fiber manufacturing process directly from PTFE emulsion, rather simpler than the processes described above, has been developed.[26] Thick strand of PTFE was directly prepared from emulsion in water, and thinning was obtained by a tensile die-drawing process at a low temperature (~90°C).

Since copolymerization is a promising way to improve physico-chemical membrane properties, several PTFE copolymers have been developed: perfluorosulfonic acid (PFSA), poly(tetrafluoroethylene-co-perfluoropropyl vinylether) (PFA), poly(tetrafluoroethylene-co-hexafluoropropylene) (FEP), and poly(ethylene-alt-tetrafluoroethylene) (ETFE) are important commercial copolymers presently used in membrane manufacturing processes.

3.2.3 Poly(ethylene chlorotrifluoroethylene)

ECTFE, mainly supplied under the brand Halar®, is a novel material with strong potential in membrane separation processes to its high resistance to solvents and alkaline solutions.[27]

ECTFE is a 1:1 alternating copolymer of ethylene and chlorotrifluoroethylene. The chemical structure of ECTFE provides excellent chemical stability, wear resistance, and high mechanical strength. Its remarkable stability in a wide range of pH (from 2 to 13) and its limited solubility in organic solvents make it suitable for membrane applications. ECTFE is a semi-crystalline, thermoplastic copolymer with a brittleness temperature lower than $-76°C$ and a melting point comprised between $200°C$ and $260°C$, depending on the molecular weight, and a continuous service temperature higher than $150°C$. As a result, it exhibits excellent mechanical properties in a wide range of temperatures and shows high durability in ambient and subambient conditions.[28] It is resistant to a variety of corrosive chemicals and organic solvents, including strong acids, chlorine, caustic solutions, strong polar solvents, and strong oxidizing agents. Due to its better processability with respect to PTFE, ECTFE is receiving increasing attention.[29] The high hydrophobicity of ECTFE membranes makes them suitable for applications in microfiltration (MF) and ultrafiltration (UF),[31,32] MD, MCr, PV.[29] This copolymer has been studied primarily in the preparation of flat sheet[30,31] and hollow fiber[32] membranes. Due to its high stability, the fabrication of ECTFE membranes is not feasible via wet- or dry-casting. A modified compression molding process has been tested in order to obtain a semi-permeable porous membrane via a plasticization step and subsequent extraction of the plasticizer with a specific solvent.[33] At elevated temperatures, ECTFE becomes soluble in selected solvents, and membranes can be prepared by TIPS process.[30,31] The preparation of ECTFE membranes by TIPS dates back to a patent[34] registered in 1987. ECTFE hollow fibers were fabricated by extruding and melt-molding with oligomer CTFE and silica particles as a plasticizer and additive, respectively. Then, the oligomer was extracted by 1,1,1-trichloroethane, and silica particles were leached by treatment with a caustic solution. However, this procedure is complicated and expensive due to the high cost of the oligomer CTFE and the necessity to extract it using a toxic halogenated solvent.

DBP can be suitably used as a diluent for ECTFE membrane preparation via TIPS due to its high boiling point and strong ability to dissolve ECTFE.[30] Since DBP is toxic, triethyl citrate (CTF) and GTA have been tested as alternative environmentally friendly diluents.[35] Recently, NMP was adopted as a solvent, whereas GTA, CTF, dibutyl itaconate (DBI), and diethyl adipate (DEA) were selected as plasticizer additives.[29]

The major disadvantages of currently available protocols for the preparation of ECTFE membranes are related to the use of high temperatures (over 230°C–260°C for DBP,[30,31] 230°C for GTA and CTF[35]). The use of NMP requires working temperatures below 200°C.[29] However, due to the relatively low boiling point of NMP, the high vapor pressure at temperatures close to 200°C might create problems in large-scale applications.

3.2.4 *Other fluoropolymers*

Hyflon® AD, Teflon® AF, and Cytop®, are typical hydrophobic amorphous fluoropolymers showing quite interesting properties for membrane preparation.

Hyflon® AD is produced by microemulsion free radical polymerization of tetrafluoroethylene (TFE) and 2,2,4-trifluoro-5-trifluoromethoxy-1,3-dioxole (TTD). Teflon® AF is a copolymer of perfluoro-2,2dimethyldioxole (PDD) with TFE. Cytop® is a homopolymer obtained by cyclopolymerization of perfluorobutenylvinylether (BVE), characterized by a controlled alternating structure with a glass-transition temperature (T_g) of 108°C.

All these materials are good candidates for preparing gas and vapor separation membranes.[36,37] Their high solubility in perfluorinated solvents, and the low solution viscosity, make them attractive for the preparation of membranes having different morphologies and structures.

Coating Hyflon® AD onto PVDF membranes results in a narrow pore size distribution, high porosity, low resistance to mass transfer, and good mechanical resistance. Hidrophobic Hyflon membranes are eligible for MGA, MD, and MCr applications.[38,39]

3.3 Preparation Methods

Different methodologies are in use to manufacture membranes. This section provides a short description of sintering, stretching, track-etching, and phase inversion technique, the latter being the most common method for preparing polymeric membranes at an industrial scale.

3.3.1 *Sintering*

Sintering is a simple technique for preparing both organic and inorganic membranes: a powder of particles is pressed into a film or plate and sintered just below the melting point. The process generates microporous membranes having porosity in the range of 10%–40% (up to 80% if metallic powder is used) and a rather irregular pore size distribution (Fig. 3.2). The typical pore size, determined by the particle size of sintered powder, typically ranges from $0.2\,\mu$m to $20\,\mu$m.

The main advantages of sintering are the high purity of membranes due to absence of chemical additives or solvents, and the good reproducibility under appropriate control of grain size. From a thermodynamic point of view, sintering is driven by a decrease in the surface free energy of compacted grains as a result of replacement of solid-vapour interfaces (surface energy γ_{sv}) with solid-solid interfaces (surface energy γ_{ss}). Sintering is, therefore, an irreversible process in which a free energy decrease is brought about by a decrease in both surface area (coarsening or grain growth) and surface tension ($\gamma_{ss} < \gamma_{sv}$) during densification.

Common polymeric powders sinterized are polyethylene, polytetrafluoroethylene, and PP. A typical morphology of resulting membranes is shown in Fig. 3.3.

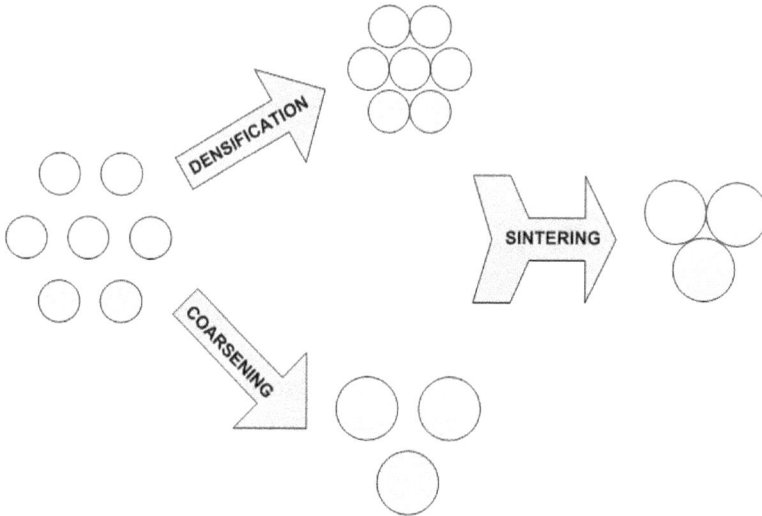

Figure 3.2. Illustration of sintering stages.[40]

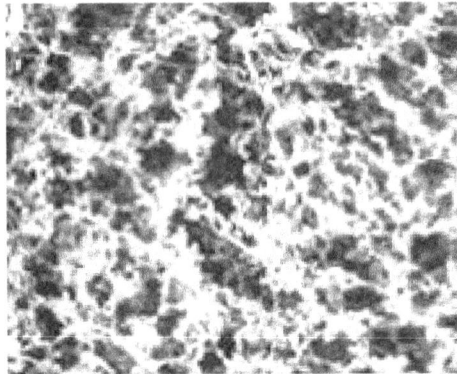

Figure 3.3. Scanning electron micrograph of a PTFE membrane prepared by sintering.

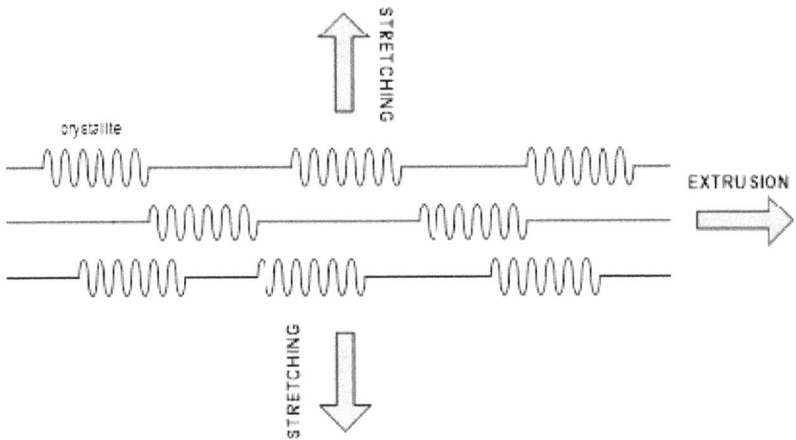

Figure 3.4. Stretching principle.

3.3.2 *Stretching*

Microporous membranes can be also prepared by stretching a homogeneous polymer film made from a partially crystalline material, obtained by extrusion of a polymeric powder at a temperature close to the melting point. Crystallites in the polymers are aligned in the direction of drawing (Fig. 3.4); after annealing and cooling, a mechanical stress is applied perpendicularly.

A typical result of tensile drawing applied to a high-density polyethylene is reported in Fig. 3.5. After a preliminary homogeneous tensile deformation

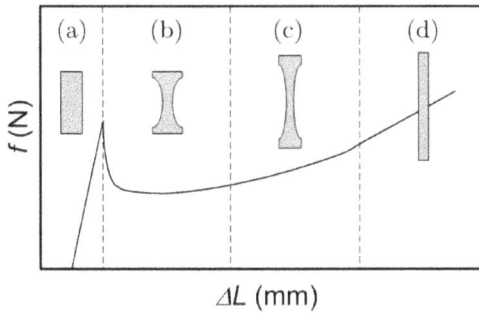

Figure 3.5. A qualitative representation of a typical tensile deformation curve of a semi-crystalline polymer.

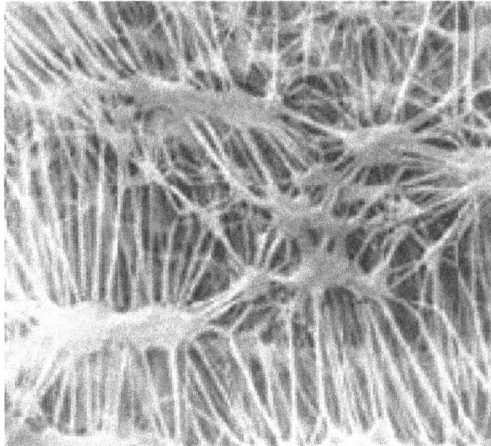

Figure 3.6. Gore-Tex PTFE membrane prepared by stretching (pore size ≈0.2 μm).

(zone a), the middle part of the specimen becomes progressively narrower (zone b); this phenomenon, called necking, often occurs for high-crystallinity or thermoplastic polymers. If the specimen is further stretched, both the neck and the neighboring parts will gradually continue to shrink until the sample reaches a certain final diameter (zone c). The last part of the diagram shows a homogeneous deformation of the whole specimen at constant diameter again (zone d).

This manufacturing process produces a relatively uniform porous structure with pore size distribution in the range of 0.2 μm–20 μm and porosity of about 90%. Typical stretched polymers are polyethylene and PTFE (Fig. 3.6).

Table 3.2.   Track-etching conditions for selected polymers.[41]

Polymer	Irradiating source	Track-etching threshold (eV/nm)	Etching solution	Etching conditions
PC	^{4}He, 0.3 MeV	250	6.25M NaOH	50°C, for minutes
PTFE	^{16}O, 36 MeV	900	50% v/v H_2SO_4 + $KMnO_4$	70°C, for hours
PI	^{16}O, 36 MeV	900	6.25M NaOH	45°C–65°C, for hours
PVDF	^{11}B, 1.2 MeV	950	5M–12M NaOH	65°C–85°C, for days

3.3.3 Track-etching

Polymeric membranes with uniform and perfectly round cylindrical pores can be obtained by track-etching. Homogeneous thin films, usually with thickness of $5\,\mu$m–$15\,\mu$m, are exposed to the irradiation of collimated charged particles, having energy of about 1 MeV. These particles damage the polymeric matrix depending on the density of irradiated energy, the polymer sensitivity to irradiation, and the operative conditions. Table 3.2 reports the track-etching threshold energy (dE/dx), representing the energy of the irradiating ions expressed in eV required to produce a cracking of 1 nm, for specific polymer/ion pairs. The film is then immersed in an acid or alkaline bath, where the polymeric material is etched away along the tracks, leaving perfect pores with a narrow size distribution.

Common track-etched polymers are polycarbonate (PC), polyethylene terefta-late (PET), PVDF, polyimide (PI), polyetheretherketone (PEEK). The membrane pore size typically ranges between $0.02\,\mu$m and $10\,\mu$m with surface porosity around 10% (Fig. 3.7).

3.3.4 Phase inversion technique

Membranes are prepared by phase inversion technique from polymers that are soluble at a certain temperature in an appropriate organic solvent or solvent mixture; the solid phase is precipitated by changing the temperature and/or composition of the system. These changes aim at creating a miscibility gap in the system at a given temperature and composition; from a thermodynamic point of view, the free energy of mixing ΔG_m of the system becomes positive.

As briefly anticipated in Section 3.2.1, the appearance of two different phases, i.e. a solid phase forming the polymeric structure (symmetric, with porosity almost uniform along the membrane cross-section, or asymmetric, with a selective thin skin layer on a sub-layer), and a liquid phase generating the pores of the membrane,

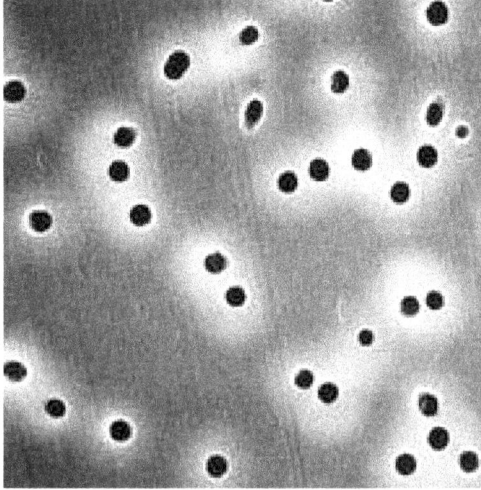

Figure 3.7.   PC membrane prepared by track-etching.

is determined by a few and conceptually simple actions:

1. changing the temperature of the system (cooling of a homogeneous polymer solution which separates in two phases): TIPS;
2. adding non-solvent or non-solvent mixture to a homogeneous solution: diffusion-induced phase separation (NIPS);
3. evaporating a volatile solvent from a homogeneous polymer solution prepared using solvents with different dissolution capacity (VIPS).

Although these procedures are operatively different, the fundamentals of membrane formation mechanisms are governed by comparable thermodynamic and kinetic concepts: variations in the chemical potential of the system, diffusivities of components in the mixture, Gibbs free energy of mixing, and formation of miscibility gaps. TIPS and NIPS processes, often also utilized in combination to prepare membranes, are discussed in detail in the following sections.

Phase separation: a thermodynamic description

The Gibbs free energy of a system is defined as a state function of enthalpy (H) and entropy (S):

$$G = H - TS, \qquad (3.1)$$

where T is the temperature of the system. In general, G depends on temperature, pressure (P), and number of moles (n_i) of components in the system:

$$G = G(T, P, n_1, n_2, \ldots, n_k).$$ (3.2)

The change in Gibbs free energy for a multi-component systems is given by

$$dG = \left(\frac{\partial G}{\partial T}\right)_{P,n_i} dT + \left(\frac{\partial G}{\partial P}\right)_{T,n_i} dP + \sum_{i=1}^{k} \left(\frac{\partial G}{\partial n_i}\right)_{T,P,n_j} dn_i$$ (3.3)

In Eq. (3.3),

$$\left(\frac{\partial G}{\partial T}\right)_{P,n_i} = -S \quad \left(\frac{\partial G}{\partial P}\right)_{T,n_i} = V \quad \left(\frac{\partial G}{\partial n_i}\right)_{T,P,n_j} = \mu_i,$$ (3.4)

and, therefore

$$dG = -SdT + VdP + \sum_{i=1}^{k} \mu_i dn_i.$$ (3.5)

In Eq. (3.4), V is the volume and μ_i is the chemical potential of the ith component. For a two-component mixture, T and P being constant, the Gibbs free energy per mole G_m is given by the sum of the chemical potentials of both components 1 and 2:

$$G_m = x_1\mu_1 + x_2\mu_2.$$ (3.6)

When n_1 moles of component 1 are mixed to n_2 moles of component 2, the change in the free energy of mixing ΔG_m per mole of mixture is

$$\Delta G_m = x_1\Delta\mu_1 + x_2\Delta\mu_2.$$ (3.7)

For an ideal solution, the chemical potential of each component is expressed by:

$$\mu_i = \mu_i^0 + RT \ln x_i,$$ (3.8)

where μ_i^0 is the molar free energy of pure components. This circumstance is graphically illustrated in Fig. 3.8.

From Eq. (3.8),

$$\Delta\mu_i = RT \ln x_i$$ (3.9)

and

$$\Delta G_m = RT (x_1 \ln x_1 + x_2 \ln x_2).$$ (3.10)

Since $\ln x_i$ is negative (because $x_i < 1$), $\Delta G_m < 0$: an ideal mixture always mixes spontaneously. Unfortunately, real mixtures do not exhibit such a simple

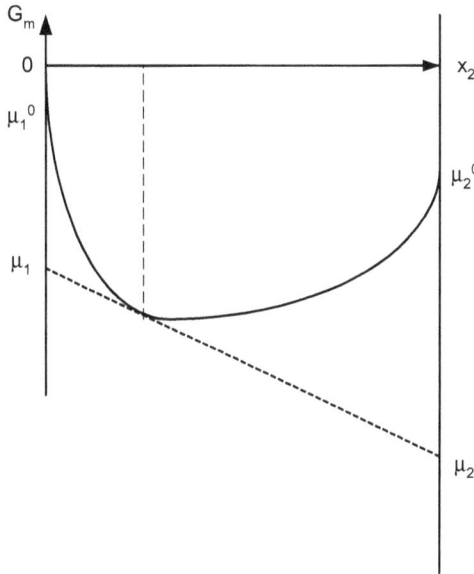

Figure 3.8.   Gibbs free energy of mixing for a two-component system at constant T and P.

behavior, and the Gibbs free energy of mixing can be conveniently expressed in terms of the enthalpy of mixing ΔH_m and the entropy of mixing ΔS_m:

$$\Delta G_m = \Delta H_m - T \Delta S_m. \tag{3.11}$$

For polymer/solvent systems, Hildebrand derived an expression for ΔH_m valid for small apolar solvents[42]:

$$\Delta H_m = V_m \left(\sqrt{\frac{\Delta E_1}{V_1}} - \sqrt{\frac{\Delta E_2}{V_2}} \right)^2 \phi_1 \phi_2, \tag{3.12}$$

where V_m, V_1, and V_2 are the molar volumes of the solution and of the two components respectively, ΔE the energy of vaporization, and ϕ the volume fraction.

The square root of the cohesive energy density $\Delta E / V$ defines the solubility parameter δ. The cohesive energy per unit volume is the energy necessary to remove a molecule from its neighboring molecules to infinite distance. For the polymer/solvent system, if $\delta_1 \approx \delta_2$, then the value of ΔH_m tends to zero and the polymer is miscible in the solvent.

The solubility parameter (expressed in SI units as $MPa^{1/2}$, equivalent to $J^{1/2} cm^{-3/2}$ or, in cgs system, to $cal^{1/2} \cdot cm^{-3/2}$, where $1\ cal^{1/2} \cdot cm^{-3/2} = 2.0455$

$MPa^{1/2}$) consists of three contributions[43]:

$$\delta^2 = \delta_d^2 + \delta_p^2 + \delta_h^2, \tag{3.13}$$

where δ_d is the solubility parameter due to dispersion forces, δ_p is the solubility parameter due to polar forces, and δ_h is the solubility parameter due to hydrogen bonding.

For an ideal gas, the Hildebrand solubility parameter is equal to the square root of the heat of vaporization ΔH_v subtracted by the term RT (where R is the universal gas constant) and divided by its molar volume:

$$\delta = \sqrt{\frac{\Delta H_v - RT}{V_m}}. \tag{3.14}$$

Predictions based on this assumption can be extended to non-polar or slightly polar solvents. The heat of vaporization ΔH_v is generally known at a given temperature (i.e. T_1); using the thermodynamic concept of critical temperature (T_c) and reduced temperature ($T_r = T/T_c$), it is possible to compute the value of ΔH_v at T_2:

$$\frac{\Delta H_v(T_2)}{\Delta H_v(T_1)} = \frac{(1 - T_{r2})^{0.38}}{(1 - T_{r1})^{0.38}}. \tag{3.15}$$

The parameter δ can be also evaluated from the constant a (expressed in units of $l^2 \cdot atm$) in the van der Waals equation:

$$\left(P + \frac{a}{V_m^2}\right)(V_m - b) = RT, \tag{3.16}$$

with

$$\delta = 1.2\frac{a^{0.5}}{V_m}. \tag{3.17}$$

Values of Hildebrand solubility parameters for solvents and polymers commonly used in the preparation of membranes are listed in Tables 3.3 and 3.4, respectively.

The entropy of mixing ΔS_m can be evaluated using the lattice model proposed by Flory and Huggins.[47] Referring to Fig. 3.9, it is assumed that macromolecules consists of segments identical in size, each occupying a site. If ω is the number of segment of a macromolecule, n_2 the number of polymeric chains, and n_1 the number of solvent molecules, then the total number of molecules N is

$$N = n_1 + \omega n_2. \tag{3.18}$$

Statistical considerations suggest that the total number of possible combinations Ω to arrange all the molecules in the lattice is given by

$$\Omega = \frac{\Gamma^{n_2}(\Gamma - 1)^{n_2(\varpi - 2)}}{n_2! \sigma^{n_2}} N^{-(\varpi - 1)n_2}\left[\varpi^{n_2}\frac{\left(\frac{N}{\varpi}\right)!}{\left(\frac{n_1}{\varpi}\right)!}\right]^{\varpi}, \tag{3.19}$$

Table 3.3. Solubility parameters of some solvents (expressed in $MPa^{1/2}$).[44]

Solvent	δ_d	δ_p	δ_h	δ
DMA	16.8	11.5	10.2	22.7
DMF	17.4	13.7	11.3	24.8
DMSO	18.4	16.4	10.2	26.7
HMPA	18.4	8.6	11.3	23.2
NMP	18.0	12.3	7.2	22.9
TMU	16.8	11.5	9.2	22.3
TEP	16.8	16.0	10.2	22.3

Table 3.4. Solubility parameters of selected polymers (*expressed in $MPa^{1/2}$;[44] **expressed in cal/cm^3;[45] ***expressed in $MPa^{1/2}$ from group contribution method).[46]

Polymer	δ_d	δ_p	δ_h	δ
Polyvinylidene fluoride*	17.2	12.5	9.2	23.2
Polypropylene*	16.6	−2.3	1.0	16.8
Polydimethylsiloxane*	—	—	—	—
% swell: 300	16.0	0.1	4.7	16.7
% swell: 400	16.0	−0.04	3.5	16.4
Polyethylene**	8.6	0	0	8.6
Nylon 66**	9.1	2.5	6.0	11.6
Polysulfone**	9.0	2.3	2.7	9.6
Polyacrylonitrile**	8.9	7.9	3.3	12.3
Cellulose acetate**	7.9	3.5	6.3	10.7
Poly(phenylene oxide)**	9.4	1.3	2.4	9.8
Polytetrafluoroethylene***	—	—	—	12.7

where σ is a constant ($= 2$ for asymmetric macromolecules), and Γ is the degree of coordination ($= 4$ for a bi-dimensional lattice).

Using the Boltzmann equation ($S = k ln \Omega$) to calculate the entropy of the solution, and subtracting the entropy of the single components, the entropy of mixing ΔS_m results:

$$\Delta S_m = -R(n_1 \ln \phi_1 + n_2 \ln \phi_2), \tag{3.20}$$

where

$$\phi_1 = \frac{n_1}{n_1 + \varpi n_2} \quad \text{and} \quad \phi_2 = \frac{\varpi n_2}{n_1 + \varpi n_2}. \tag{3.21}$$

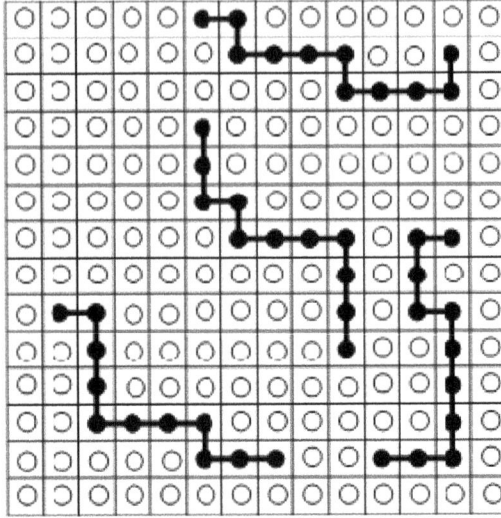

Figure 3.9.   A lattice representation of a polymer-solvent mixture.

Under the hypothesis that $\Delta H_m = 0$, the Gibbs energy of mixing is

$$\frac{\Delta G_m}{NRT} = -\frac{\Delta S_m}{R} = \left(\frac{\phi_1}{n_1}\right)\ln\phi_1 + \left(\frac{\phi_2}{n_2}\right)\ln\phi_2. \qquad (3.22)$$

In this case, it can be shown that athermal polymer solutions never demix. Demixing occurs in presence of a positive enthalpy term $\Delta H_m > 0$. In the most general case of $\Delta H_m \neq 0$, the expression for Gibbs free energy of mixing is generalized by including an enthalpic contribution $(n_1\phi_2\chi)$:

$$\Delta G_m = RT(n_1 \ln\phi_1 + n_2 \ln\phi_2 + n_1\phi_2\chi), \qquad (3.23)$$

where χ is the Flory–Huggins interaction parameter.

Figure 3.10 shows diagrams of ΔG_m vs. ϕ and its derivative up to the second order at a given temperature and pressure for a bi-component system. The mixture is stable over a certain composition range if

$$\frac{\partial G_m}{\partial \phi} < 0. \qquad (3.24)$$

The first derivative becomes zero when the binodal curve is reached, and attains a positive maximum when the spinodal curve is reached. In this region, the mixture is metastabile: there is not a driving force to a spontaneous demixing.

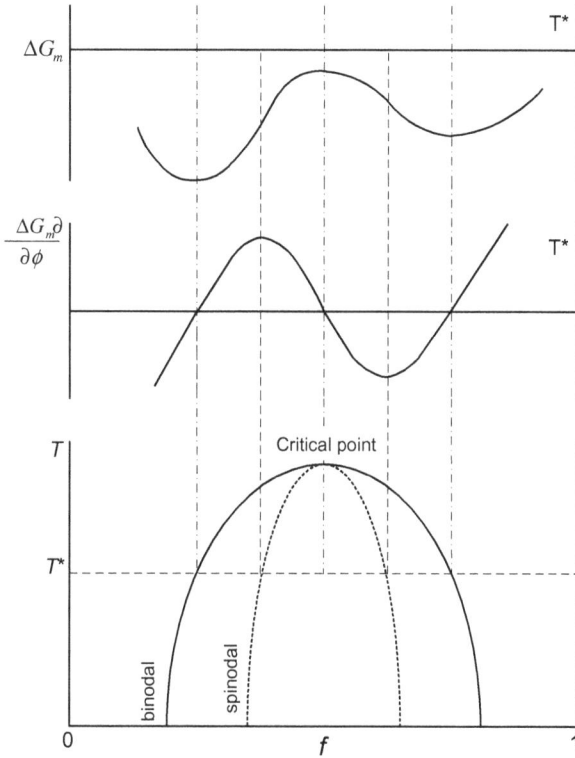

Figure 3.10. Plot of the Gibbs free energy of mixing as a function of the composition, referred to as a system exhibiting a miscibility gap.

A thermodynamic instability is observed when

$$\frac{\partial^2 G_m}{\partial \phi^2} < 0, \tag{3.25}$$

and the system will demix spontaneously. The critical point is individuated by a zero value of the first derivative and a minimum in the second derivative.

Non-solvent induced phase separation (NIPS)

In the NIPS process, the membrane is formed by polymer precipitation caused by concentration variations and diffusive interchange between the solvent and the non-solvent. The final structure of the membrane is determined by the rate of polymer precipitation. A low precipitation rate results in the formation of a symmetric structure, whereas a high precipitation rate leads to an asymmetric membrane, with large voids, spongy sublayer, and/or finger-like cavities below a microporous or dense upper layer.

From a practical point of view, NIPS process is routinely used to prepare integral asymmetric membranes. This process is articulated in few simple steps:

1. dissolution of a polymer in an appropriate solvent to form a solution typically containing 10 wt.%–30 wt.% of polymer;
2. casting the solution on a suitable support into a film of $100\,\mu m$–$500\,\mu m$ thickness;
3. quenching of the film in the non-solvent (typically water or an aqueous solution), or evaporation of the solvent to increase the polymer concentration.

During the third step, the homogeneous polymeric solution demixes into two phases: a polymer-rich solid phase, which forms the structure of the membrane, and a solvent-rich liquid phase, which results in the formation of liquid-filled membrane pores.

For a more quantitative explication of the NIPS process, let us consider the equilibrium diagram of the ternary system polymer/solvent/non-solvent reported in Fig. 3.11. The three-component mixture exhibits a miscibility gap in a certain composition range defined by the binodal curve; under these conditions, the system splits into two distinct phases. Starting from a homogeneous mixture (point A), if the solvent is completely removed, the final composition of the system will be represented by point B located on the polymer/non-solvent axis. Here, the system is distributed in two phases, whose compositions are indicated by point B' (polymer-rich phase, solid membrane structure) and point B'' (non-solvent rich phase, liquid-filled pores). Liquid–liquid demixing occurs when the line A-B intersects the binodal curve.

Thermodynamic information about the system is useful but not sufficient to predict the resulting morphology of the membrane. A complete understanding of membrane formation is difficult because of the high number of involved mechanisms and phenomena, i.e. thermal effects, demixing kinetics, eventual presence of additives, mutual interaction parameters between polymer/solvent/non-solvent, ternary diffusivities, initial and boundary conditions etc.

As a practical case, cloud point data at 20°C for PVDF-solvent-water systems are illustrated in the ternary phase diagram shown in Fig. 3.12[45]; at a fixed concentration of polymer, the amount of water required to precipitate PVDF increases according to the following order for different solvents:

HMPA > DMA > DMF > TEP > TMP.

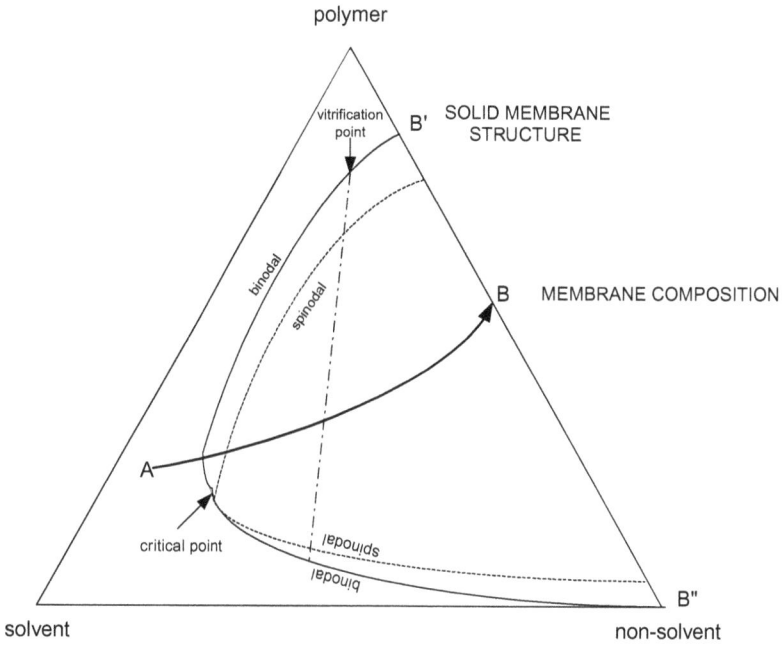

Figure 3.11. A three-component system isothermal diagram showing the formation of a membrane by solvent evaporation. (Reprinted from [1] with permission. Copyright © 2005 Elsevier B.V.).

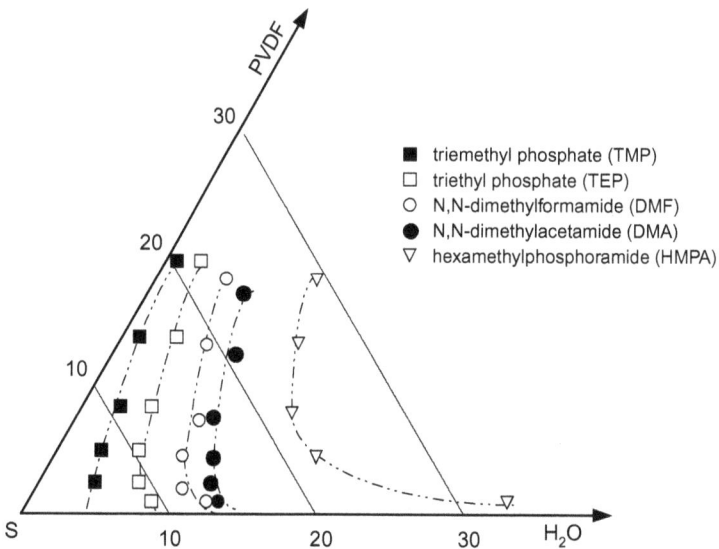

Figure 3.12. Ternary phase diagram at 20°C for a PVDF-solvent-water system. (Reprinted from [1] with permission. Copyright © 2005 Elsevier B.V.).

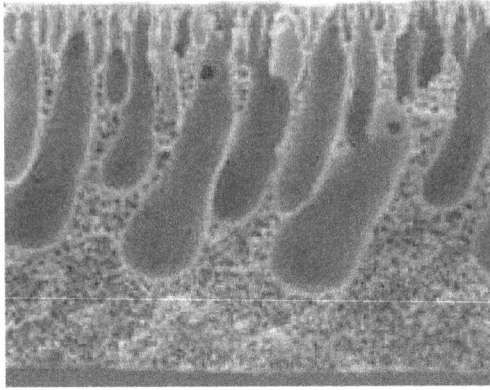

Figure 3.13.   Cross section of a PVDF membrane prepared by immersion precipitation.

A typical asymmetric structure of PVDF membrane prepared by NIPS technique is illustrated in Fig. 3.13.

The morphology of the membrane is determined by the peculiar physicochemical properties of the system. Polymer-solvent interactions have been widely investigated by various authors.[48–50] In general, low interactions correspond to a higher rate of polymer precipitation, thus resulting in the formation of finger-like structures. The compatibility of polymer and solvent can be evaluated in terms of the three components of the solubility parameter δ (see Eq. (3.13)):

$$\delta_{P-S} = \sqrt{(\delta_{d,P} - \delta_{d,S})^2 + (\delta_{p,P} - \delta_{p,S})^2 + (\delta_{h,P} - \delta_{h,S})^2}, \qquad (3.26)$$

where subscripts P and S indicate the polymer and the solvent, respectively.

The tendency of a solvent to mix with the non-solvent also affects the membrane porosity and structure.[50–54] For asymmetric membranes prepared by immersion in water, a high difference of the solubility parameter of solvent and water ($\delta_W = 47.8\,MPa^{1/2}$),[44] corresponding to a low tendency to mix, results in a high membrane porosity.

A low solution viscosity generally determines the occurrence of cavities in the membrane. On the other hand, an increase of the solution viscosity due to an increase of polymer concentration obstructs the penetration of the non-solvent during the immersion step.

The rate of phase separation depends on the degree of penetration in the demixing gap. An instantaneous liquid–liquid demixing results in the formation of porous membranes; conversely, when a delay in liquid–liquid demixing occurs, dense membranes are produced. Particularly during the first instants subsequent to the immersion of the casting solution in the precipitation bath, mass transfer (solvent and non-solvent interdiffusion) could become the controlling mechanism for skin formation. It has been estimated that, for a solvent-non-solvent diffusivity on the order of 10^{-5} cm^2/s and a typical skin thickness of about 0.1 μm, the characteristic time for mutual diffusion t_d is of about 10^{-5} s.[55] Modeling studies report that, when the value of the solvent/non-solvent diffusivity increases, the concentration path in the ternary diagram should lead to entry into the demixing gap at higher polymer concentration.[56]

Thermally induced phase separation (TIPS)

TIPS results in a solid–liquid phase separation by removing thermal energy from the system. The TIPS process basically consists of four steps[57]:

1. preparation of a homogeneous solution by melt-bending the polymer with a high-boiling, low-molecular weight diluent;
2. solution casting;
3. cooling to induce phase separation and solidification of the polymer;
4. removal of the diluent (typically by solvent extraction) to form the membrane structure.

TIPS can be applied to a wide range of polymers whenever their reduced solubility in organic solvents prevents the use of non-solvent induced phase inversion. This preparation technique allows one to obtain isotropic microporous structures.

The formation of a membrane can be explained by referring to appropriate equilibrium phase diagrams, and to the theory of phase equilibria in polymer systems. For binary polymer-solvent systems in which the polymer is semi-crystalline, the melting point of the polymeric compound is related to the mixture composition as follows[4]:

$$\frac{1}{T_m} - \frac{1}{T_m^0} = \frac{R}{\Delta H_f} \frac{V_2}{V_1} (\phi_2 - \chi \, \phi_2^2), \tag{3.27}$$

where T_m and T_m^0 are the melting temperatures of the crystalline polymer in solution and the pure crystalline polymer, respectively; V_1 is the molar volume of the solvent, V_2 is the molar volume for the repeating unit, ϕ_2 is the volume fraction of the solvent, and χ is the Flory–Huggins interaction parameter.

Solving Eq. (3.27) for T_m yields

$$T_m = \frac{1}{\frac{R}{\Delta H_f} \frac{V_2}{V_1} (\phi_2 - \chi \phi_2^2) + \frac{1}{T_m^0}}, \tag{3.28}$$

and plotting it as function of the volume fraction of the polymer, $\phi_1 = (1 - \phi_2)$, it is possible to derive a temperature-composition diagram for a semi-crystalline polymer-diluent system.

The temperature at which phase separation occurs is increased if the strength of interactions in the polymer-solvent is lowered (χ increases). Equation (3.28) also shows that, all parameters being constant, the smaller the molar volume of the solvent with respect to the molar volume of the polymer, the larger the melting point depression.

Referring to Fig. 3.14, let us consider an homogeneous polymer-solvent solution (point A) at temperature T_A. If the solution is progressively cooled at the same composition, the system loses its stability and a solid-liquid separation occurs. At point B (temperature T_B) the system separates in a polymer-rich phase — having conposition indicated by the point B″ — and in a solvent-rich phase — represented by the point B′. According to the lever rule, segments B′ − B and B − B″ represent the ratio of the amounts of the two phases in the mixtures, from which it is possible to roughly estimate the porosity of the membrane.

The membrane structure originates from the polymer-rich phase, while the solvent-rich phase forms the liquid filled pores. Table 3.5 lists some physicochemical properties of polymeric materials usually employed in TIPS process.

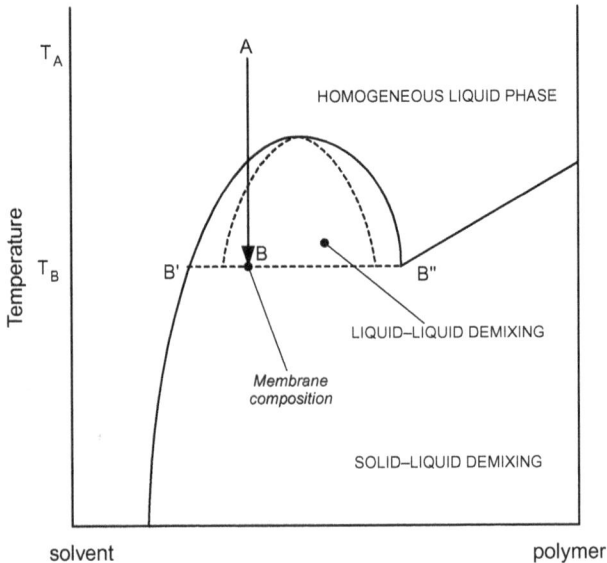

Figure 3.14. Phase diagram for a polymer-solvent binary system as a function of temperature. (Reprinted from [1] with permission. Copyright © 2005 Elsevier B.V.)

Table 3.5. Polymers for membrane prepared by TIPS.[57]

Polymers	Density (g/cm^3)	Average molecular weight (Da)	Melting point (°C)
Polypropylene (Himont Pro-fax 6723)	0.903	243,000	176
High density polyethylene (American Hoechst Hostalen GM-9255-F2)	0.954	224,000	130
Polychlorotrifluoroethylene, (Kel-F, 3M Company, Grade 6300)	2.050	N.A.	197
Poly (4-methyl-1-pentene) (Mitsui Chemicals, Grade RT-18)	0.835	N.A.	230
Poly(vinylidene fluoride) (Soltex Solvey 1011)	1.780	N.A.	169
Diluent			**Initial boiling point (°C)**
Mineral oil (Plough Inc., Nujol)	0.866	N.A.	~320
Kel-F oligomineral oil (3M Company, KF-3)	1.930	630	270
Dibutyl phthalate (Aldrich Chemicals)	1.043	278.4	340

A diagram of solid–liquid phase separation for polypropylene-mineral oil at different polymer concentrations is reported in Fig. 3.15. Crystallization curves are affected by the kinetics of cooling. Results show that the temperature of demixing significantly decreases if the cooling rate is increased A scanning electron micrograph of a microporous polypropylene membrane prepared by TIPS is shown in Fig. 3.16.

3.4 Membrane Modification

A large part of the commercial microporous polymeric membranes available in capillary and flat-sheet forms, at present in use for membrane contactor applications, were originally manufactured and optimized for MF purposes. The preparation of new membranes for specific operations (such as membrane distillation/crystallization) has recently increased in interest, and significant results reached in the preparation and modification methods have contributed to increased reliability of membrane contactors technology.

3.4.1 *Additives in the casting solution*

The use of additives in the casting solution, e.g. in the form of water-soluble polymers such as polyvinyl pirrolidone (PVP), polyethylene glycol (PEG), or

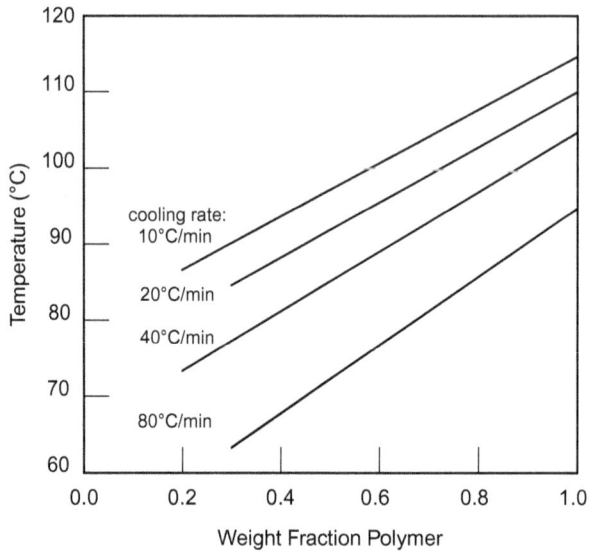

Figure 3.15. Crystallization temperature-concentration curves for PP-mineral oil at cooling rates indicated in the diagram. (Reprinted from [1] with permission. Copyright © 2005 Elsevier B.V.)

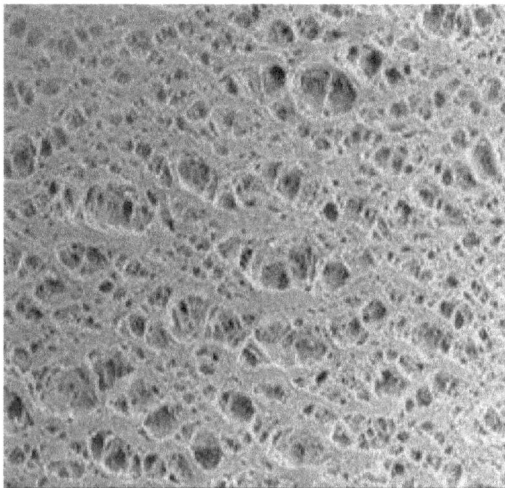

Figure 3.16. Microporous polypropylene membrane produced by thermally induced phase separation technique.

inorganic salts (LiCl), is a simple and practical way to modulate the membrane morphology. This aspect has been investigated in the preparation of microporous PVDF membranes for MD applications, where high porosity is requested in order to obtain a significant flux.[58–60] In particular, it has been observed that the addition of LiCl increases the rate of PVDF precipitation during the immersion step, which causes the formation of an open structure with large macrovoids and cavities. The fast precipitation is due to both a high tendency of the additive to mix with water, and to interactions of the additive with polymer and solvent.[61] The effect of LiCl content on the porosity and mean pore size of membranes prepared from DMA:PVDF = 88:12 is illustrated in Fig. 3.17. Porosity progressively increases from 79% to 83% in the range of 1 wt.%–7 wt.% of additive, while the mean pore size achieves a maximum of 0.04 μm corresponding to LiCl 3.5 wt.%.

3.4.2 Use of copolymers

Membranes can be prepared by using simultaneously different types of monomers. Asymmetric and composite membranes showing a high hydrophobic character and contact angles to water greater than 120° have been prepared from copolymers of TFE and TTD, commercially known as HYFLON AD.[62]

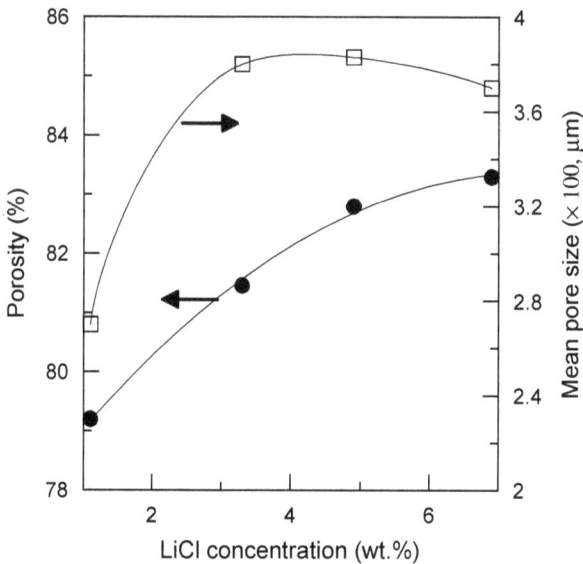

Figure 3.17.   Effect of LiCl on the porosity and mean pore size of a PVDF membrane using DMA as a solvent. Reprinted from [1] with permission. Copyright © 2005 Elsevier B.V.)

Asymmetric hydrophobic membranes from a copolymer of TFE and VDF have been prepared by phase inversion process.[60] According to the experimental analysis, these membranes exhibit excellent mechanical properties (stretching strain and extension ratio at break approximately 6–8 times higher PVDF) and good hydrophobicity (contact angle to water of about 87°).

3.4.3 *Composite membranes*

Composite membranes generally exhibit an asymmetric structure, generated by the deposition of a thin top layer on a porous sub-layer of different material. Composite membranes have the advantage that the properties of each layer can be independently modulated and optimized in order to obtain the required persormance in terms of selectivity, permeability, chemical and thermal stability, etc.

Several preparation procedures for composite membranes can be identified; most common are:

(1) casting the thin layer separately (e.g. by spreading a very dilute polymer solution on the surface of a water bath) and then laminating it on a microporous support;
(2) coating the microporous support by a polymer, a reactive monomer or a pre-polymer solution (e.g. by immersion in an appropriate solution with low solute concentration — often less than 1%), followed by drying, heat treatment or radiation;
(3) plasma polymerization;
(4) interfacial polymerization of reactive monomers of the surface of the microporous support.

Details about each of these techniques can be found elsewhere.[48] Some specific examples related to the preparation of composite membranes for membrane contactors applications are shortly mentioned below.

The work of Xu and colleagues[63] showed that hydrophobic PTFE membranes with a protective hydrophilic sodium alginate coating were resistant to wet-out at least for 300 minutes during osmotic distillation tests using orange oil feed up to 0.8 wt.%. The reduction in the overall mass transfer coefficient due to the coating was less than 5%.

Hydrophilic/hydrophobic composite membrane were prepared Wu *et al.*[64] treating a hydrophilic porous cellulose acetate surface via radiation graft polymerization of styrene.

Low pressure plasma polymerization makes possible to apply a thin layer upon a porous sub-layer; this generally results in a change of the chemical composition and properties of a material, such as wettability, refractive index, hardness, etc.

Plasma is obtained by ionising a gas by high frequency (up to 10 MHz) electrical discharges. The pressure inside the reactor usually varies between 0.1 mbar and 10 mbar. High hydrophobicity, somewhat higher than that of PTFE, was achieved by fluorinated coatings named "Teflon-like".[65] Kong and coworkers[66] have modified hydrophilic microporous cellulose nitrate membranes by plasma polymerization of octafluorocyclobutane. The performance of these membranes, tested in membrane distillation applications, was found comparable with that of usual hydrophobic polymers.

3.4.4 *Surface modifying molecules*

In general, an increase of membrane porosity and pore size enhances the trans-membrane flux. The analogous effect can be obtained if membrane thickness and tortuosity is decreased. When considering thermally-driven membrane contactor operations, such as in the case of membrane distillation/crystallization, the conductive heat loss increases when using thin membranes; as a consequence, the thermal efficiency of the process is reduced. The preparation of composite microporous hydrophobic/hydrophilic membranes is a viable option to resolve the conflict between the requirements for high mass transfer and low heat transfer through the membrane; specifically, the top hydrophobic thin layer is responsible for the mass transport, while the hydrophilic sub-layer increases the resistance to the conductive heat flux. Khayet *et al.*[67] have modified the surface of hydrophilic membranes by adding oligomeric fluoropolymers synthesized by polyurethane chemistry and tailored with fluorinated end-groups. During membrane formation, surface-modifying molecules (SMMs) migrate to the air-film surface according to the thermodynamic tendency to minimize the interfacial energy. These modified membranes exhibit low surface energies, good mechanical strength, and high chemical resistance.[68]

3.5 Hydrophobic/Hydrophilic Properties, Contact Angle, Surface Tension

The characterization of the surface chemistry is a critical issue in membrane crystallization technology, since the nucleation kinetics strongly depends on the hydrophobic/hydrophilic character of the polymer layer, surface charge, interactions between membrane and solutes or solvents, etc.

3.5.1 *Contact angle*

Contact angle measurement is a traditional method for describing the hydrophobic or hydrophilic behavior of a material. In principle, it provides information about the wettability of an ideal surface; in most practical applications, the intrinsic value of the contact angle is perturbed by surface porosity and roughness, heterogeneity, etc.

The contact angle made by a liquid droplet deposited on a smooth surface is greater than 90° if the affinity between liquid and solid is relatively low; in the case of water, the material is considered hydrophobic. Hydrophilc membranes are defined for a water contact angle below 90°. Ideally, wetting occurs at 0°, when the liquid spreads onto the surface.

Referring to Fig. 3.18, at the triple point C (where solid-liquid-vapor interfaces are in contact) the mechanical equilibrium is expressed by the Young equation:

$$\gamma_{LV} \cos\theta = \gamma_{SV} - \gamma_{SL}, \tag{3.29}$$

where γ_{LV}, γ_{SV}, and γ_{SL} are the surface tension for liquid-vapor, the surface energy of the polymer, and the solid–liquid surface tension, respectively. Surface tension values γ_{LV} for different test liquids are reported in Table 3.6.

Surface tensions involving a solid cannot be measured directly; therefore, a second equation is required to determine the hydrophobicity of the material.

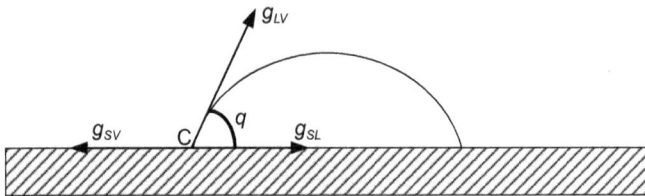

Figure 3.18. Contact angle (θ) of a liquid droplet deposited on the surface of a solid; representation of the thermodynamic equilibrium at the triple point C.

Table 3.6. Surface tension γ_{LV} for different test liquids.

Test liquid	γ_{LV}(mJ/m)
Water	72.8
Glycerol	64
Formamide	58
Diiodomethane	50.8
Ethylene glycol	48
A-bromonaphthalene	44.4
Dimethylsulfoxide	44
Chloroform	27.2

Using a thermodynamic approach, Newmann[69] established the following equation of state to relate the three interfacial tensions

$$\gamma_{SL} = \gamma_{SV} + \gamma_{LV} - 2\sqrt{\gamma_{SV}\gamma_{LV}}\exp[-\beta(\gamma_{SV} - \gamma_{LV})^2], \qquad (3.30)$$

and, combining it with the Young equation,

$$\cos\theta = -1 + 2\sqrt{\gamma_{SV}/\gamma_{LV}}\exp[-\beta(\gamma_{SV} - \gamma_{LV})^2], \qquad (3.31)$$

where β is a parameter independent of the solid and liquid used.

The Young equation is rigorously applicable if the solid substrate is smooth and if the surface is homogeneous and rigid, chemically inert, and insoluble to contacting liquids. The effect of surface heterogeneity on contact angle is generally established by Eq. (3.32), which predicts the contact angle θ^* of a rough surface from the contact angle θ of the equivalent smooth surface[70]:

$$\cos\theta^* = f_1\cos\theta - f_2, \qquad (3.32)$$

where f_1 and f_2 are the fractions of liquid–solid and liquid–air surfaces, respectively.

For porous membranes, Courel et al.[71] derived an expression similar to Eq. (3.32), where $f_1 = y$ and $f_2 = 1 - y$, y being the fraction of membrane surface made of solid material. For membranes with surface porosity lower than 0.5, it is generally assumed that $1 - y = \varepsilon/\tau$, where ε is the porosity and τ the pore tortuosity. For PTFE membranes, a more specific model has been developed:

$$\cos\theta* = y^2\cos\theta - (1 - y)^2 - 2y(1 - y)\sqrt{\frac{\gamma_{SV}}{\gamma_{LV}} - \cos\theta}. \qquad (3.33)$$

Under the assumption that the contact angles on the three-phase lines both on the outer drop border and over the pores are equal (as exemplified in Fig. 3.19), Troger and colleagues[72] have obtained a general relation between the observed contact angle θ' and the ideal one θ (to be observed on ideally smooth surface):

$$\cos\theta = \cos\theta' - \frac{4\varepsilon}{1 - \varepsilon}\frac{\cos\theta' + 1}{\cos\theta' - 1}, \qquad (3.34)$$

where ε is the porosity of the porous material; the validity of Eq. (3.34) has been tested on porous PTFE membranes with appreciable results.

3.5.2 Good–van Oss–Chaudhury method

The Good–van Oss–Chaudhury method[73] represents a more comprehensive approach to the determination of the surface tension components by contact angle

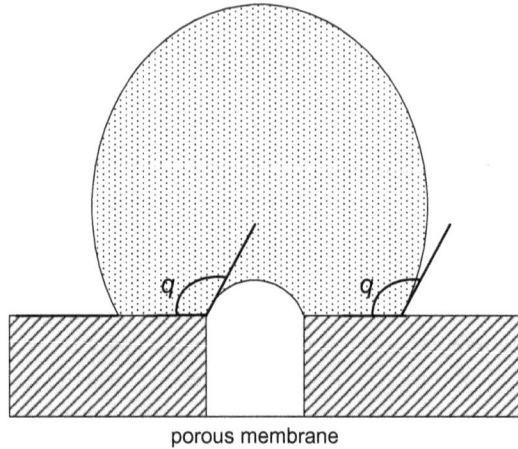

Figure 3.19. A graphical representation of the assumption that the contact angles on the three-phase lines occurring in a porous structure are equal (72).

measurements. Three reference liquids (typically: water, diiodomethane, and glycerol) are used to determine the apolar Lifshitz–van der Waals component γ^{LW}, the acid-base component γ^{AB}, the acid (electron acceptor) γ^+, and the base (electron donor) component γ^- of the surface energy.

Specifically, diiodomethane (apolar test liquid) allows at evaluating the Lifshitz–van der Waals component γ_s^{LW} of the membrane surface tension reflecting the dipole interactions:

$$\gamma_s^{LW} = \frac{\gamma_l^{LW}(1 - \cos\theta)^2}{4},\tag{3.35}$$

where subscripts s and l indicate solid (membrane) and liquid, respectively. Other components are calculated by the following equations:

$$\gamma_{LV}(1 + \cos\theta) = 2\left(\sqrt{\gamma_s^{LW}\gamma_l^{LW}} + \sqrt{\gamma_s^+\gamma_l^-} + \sqrt{\gamma_s^-\gamma_l^+}\right)\tag{3.36}$$

$$\gamma^{AB} = 2\sqrt{\gamma_s^-\gamma_s^+}.\tag{3.37}$$

Typical contact angles of water are in the range of 110–120° for PP and PTFE[74,75] and around 90° for PVDF Surface tension data for some polymeric membranes are reported in Table 3.7.

Table 3.7. Contact angles in water (W), glycerol (G) and diiodomethane (D), and surface tension parameters of different polymeric membranes: γ^{LW} = Lifshitz–van der Waals component, γ^- = electron donor component, γ^+ = electron acceptor component, and γ^{AB} = acid-base component of the liquid surface tension γ.[76]

Membrane	θ_D (°)	θ_W (°)	θ_G (°)	γ^{LW} (mJ/m²)	γ^- (mJ/m²)	γ^+ (mJ/m²)	γ^{AB} (mJ/m²)
Nylon	24 ± 2	49 ± 2	75 ± 2	47	57	4	29
Polyester	41 ± 2	75 ± 2	81 ± 1	39	15	0.9	72
Polyether-sulfone	30.4 ± 1	54 ± 1	69 ± 2	44	41	1	14
Polyurethane	35 ± 3.6	94.4 ± 1.4	100.9 ± 1.6	42.0	9.0	5.52	14.1
PEEK-WC-20	24.6 ± 2.4	69.9 ± 2.2	69.5 ± 1	46.3	15.7	0.32	4.5
Polyetherethe-rketone-WC-60	26.2 ± 2.5	71.2 ± 2.8	68.4 ± 0.9	45.7	13.7	0.15	2.9
Collagen	52 ± 1.3	82 ± 2.2	67 ± 7.6	33.4	2.3	1.1	3.2

Table 3.8. Penetration surface tension γ_L^P calculated for microporous PVDF membrane (maximum pore diameter: $0.25\,\mu$m, porosity: 75%, thickness: $200\,\mu$m) and PP membrane (maximum pore diameter: $0.40\,\mu$m, porosity: 80%, thickness: $160\,\mu$m).[70]

Aqueous solution	PVDF		PP	
	Concentration (wt.%)	γ_L^P (mN/m)	Concentration (wt.%)	γ_L^P (mN/m)
Ethanol	23.5	36.5	41.0	29.5
Acetic acid	31.0	43.4	83.0	31.8
AC	33.0	35.0	54.0	29.8
DMF	43.0	48.0	98.0	35.5

3.5.3 Contact angle and wettability

From the comparison of experimental results and theoretical calculations, Franken and colleagues[70] concluded that contact angle measurements on homogeneous smooth materials are not suitable for an accurate description of the membrane-wetting phenomenon in microporous hydrophobic membranes. Wettability criteria based on the concept of penetration surface tension γ_L^P (defined as the surface tension of the liquid on the verge of penetrating a porous medium, and measured by the penetrating drop method) seems to provide more satisfying results. Table 3.8 collects some values of γ_L^P for PVDF and PP membranes, measured using different aqueous solutions.

Breakthrough pressure

Membrane crystallization operations impose that the penetration of liquid through the micropores of the polymeric membrane must be avoided. By definition, a fluid does not pass through pores as long as the pressure is kept below a critical threshold known as "breakthrough pressure". The Laplace equation provides a relationship between the largest pore radius of the membrane $r_{p,max}$ and the breakthrough pressure ΔP_{entry}:

$$\Delta P_{entry} = -\frac{2\Theta\gamma\cos\theta}{r_{p,\max}}, \tag{3.38}$$

where γ is the interfacial tension, Θ is a geometric factor related to the pore structure (equal to 1 for cylindrical pores), and θ the contact angle.

It is reported that, for a hydrophobic membrane having a contact angle of 130°, the penetration pressure of a ideally cylindrical pore with diameter 1 μm is only 185 kPa.[77] Breakthrough pressure data for different membranes can be found in literature[78]; for most practical applications and commercially available membranes, ΔP values range between 100 kPa and 400 kPa.[79]

Breakthrough pressure is drastically reduced in presence, even at trace levels, of detergents and surfactants (because they reduce the surface tension), or organics. Experimental investigations demonstrate that, once a membrane is wetted by the penetrating liquid, a decrease in hydrostatic pressure is not able to restore the original unwetted condition.[80]

This phenomenon is qualitatively illustrated in Fig. 3.20. Liquid floods the largest pores as soon as the pressure overcomes ΔP_{entry}; when the pressure is furtherly increased, all the pores are flooded and the transmembrane flux J obeys Darcy's law:

$$J = k\Delta P, \tag{3.39}$$

k being a constant. If the system is depressurized the flux decreases linearly.

For mixtures of water and ethanol, Gostoli and Sarti[81] observed that the liquid entry pressure decreases linearly at increasing alcohol concentration.

3.6   The Influence of Pore Size Distribution

Inaccuracy when modeling the mass transfer in membrane distillation/crystallization often reflects an inadequate knowledge of the morphology of a microporous membrane.[82] A good agreement between theoretical and experimental results was obtained by Martinez-Diez and Vazquez-Gonzalez[83] using pore size distribution measured by mercury porometry and liquid displacement methods.

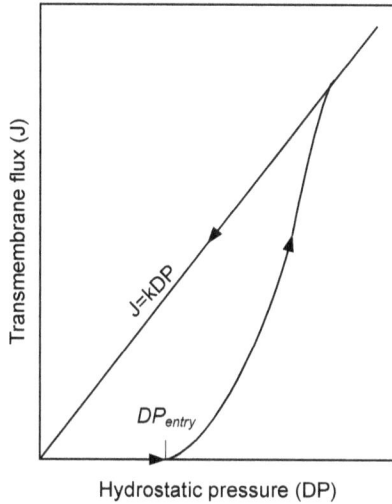

Figure 3.20. The characteristic trend of the liquid transmembrane flux versus pressure drop in microporous hydrophobic membranes.

If $f(r)$ is the normalized distribution, and $J(r)$ the transmembrane flux through pores with radius r, the total flow rate J_T through the membrane is obtained as:

$$J_T = \int_0^\infty J(r)\pi(r)^2 f(r)dr. \tag{3.40}$$

Typically, a lognormal distribution is sufficiently accurate to model the size distribution of membrane pores:

$$f(r) = -\frac{1}{SD_{\log}r\sqrt{2\pi}}\exp\left\{-\frac{1}{2}\left[\frac{\ln(r/\bar{r})}{SD_{\log}}\right]^2\right\}, \tag{3.41}$$

where $f(r)$ is the number of pores with pore radius r, $\bar{r}$ the mean pore radius, and SD_{log} the standard deviation of the lognormal function.

Figure 3.21 depicts the pore-size distribution of a PP Accurel® hydrophobic membrane, frequently used in membrane contactors experiments.

Laganà and coworkers[85] have investigated the effect of Gaussian (symmetric) and logarithmic (asymmetric) pore size distribution functions; it was found that a non-symmetrical distribution achieved a better agreement with experimental flux data. Several mathematical models aiming at inspect the influence of pore-size distribution in Membrane Distillation were tested by Phattaranawik et al.[86] Transmembrane flux predictions based on a log-normal pore size distribution were

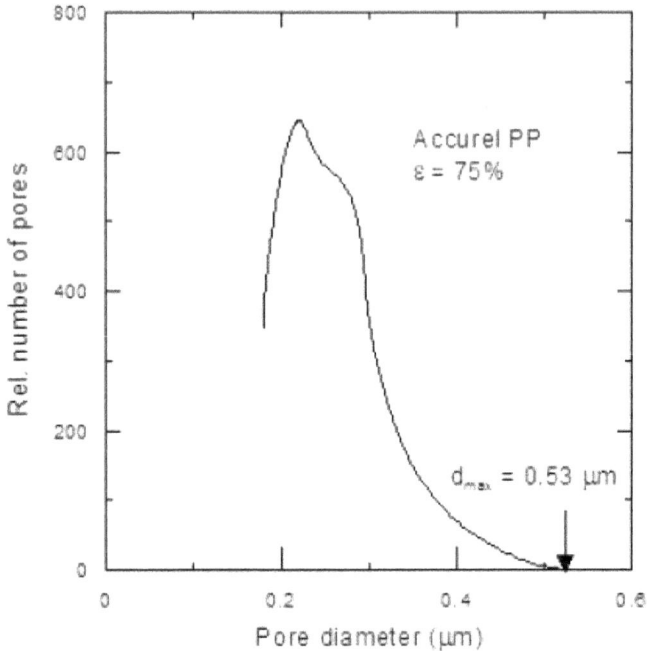

Figure 3.21. Pore-size distribution of an Accurel® PP membrane. (Reprinted from [1] with permission. Copyright © 2005 Elsevier B.V.)

in good agreement with experimental data for PVDF (Millipore, 0.22 mm) and in excellent agreement for PVDF (Millipore, 0.45 mm) and PTFE (Sartorius, 0.2 mm).

References

1. Drioli, E., Criscuoli, A. and Curcio, E. (2005). *Membrane Contactors. Fundamentals, Potentialities and Applications*, Elsevier, The Netherlands.
2. Souzy, R. and Ameduri, B. (2005). Functional fluoropolymers for fuel cell membranes, *Prog. Polym. Sci.*, **30**, 644–687.
3. Drobny, J.G. (2009). *Technology of Fluoropolymers, 2nd edn.*, CRC Press, New York, p. 227.
4. Ameduri, B. (2009). From vinylidene fluoride (VDF) to the applications of VDF-containing polymers and copolymers: Recent developments and future trends, *Chem. Rev.*, **109**, 6632–6686.
5. Dillon, D.R., Tenneti, K.K., Li, C.Y. *et al.* (2006). On the structure and morphology of polyvinylidene fluoride–nanoclay nanocomposites, *Polymer*, **47**, 1678–1688.
6. Yip, Y. and McHugh, A.J. (2006). Modeling and simulation of nonsolvent vapor-induced phase separation, *J. Membr. Sci.*, **271**, 163–176.

7. Li, C.L., Wang, D.M., Deratani, A. *et al.* (2010). Insight into the preparation of poly(vinylidene fluoride) membranes by vapor-induced phase separation, *J. Membr. Sci.*, **361**, 154–166.

8. Mu, C.X., Su, Y.L., Sun, M.P. *et al.* (2010). Fabrication of microporous membranes by a feasible freeze method, *J. Membr. Sci.*, **361**, 15–21.

9. Su, Y.L., Liang, Y.G., Mu, C.X. *et al.* (2011). Improved performance of poly(vinylidene fluoride) microfiltration membranes prepared by freeze and immersion precipitation coupling method, *Ind. Eng. Chem. Res.*, **50**, 10525–10532.

10. Feng, C., Khulbe, K.C., Matsuura, I. *et al.* (2008). Production of drinking water from saline water by air-gap membrane distillation using polyvinylidene fluoride nanofiber-membrane, *J. Membr. Sci.*, **311**, 1–6.

11. Gopalan, A.I., Santhosh, P., Manesh, K.M. *et al.* (2008). Development of electrospun PVdF-PAN membrane-based polymer electrolytes for lithium batteries, *J. Membr. Sci.*, **325**, 683–690.

12. Gopal, R., Kaur, S., Ma, Z.W. *et al.* (2006). Electrospun nanofibrous filtration membrane, *J. Membr. Sci.*, **281**, 581–586.

13. Yee, W.A., Kotaki, M., Liu, Y. *et al.* (2007). Morphology, polymorphism behavior and molecular orientation of electrospun poly(vinylidenefluoride) fibers, *Polymer*, **48**, 512–521.

14. Liu, F., Hashim, N.A., Liu, Y.T. *et al.* (2011). Progress in the production and modification of PVDF membranes, *J. Membr. Sci.*, **375**, 1–27.

15. Yang, J.A., Wang, X.L., Tian, Y. *et al.* Morphologies and crystalline forms of polyvinylidene fluoride membranes prepared in different diluents by thermally induced phase separation, *J. Polym. Sci., Part B Polym. Phys.*, **48**, 2468–2475.

16. Ameduri, B. (2009). From vinylidene fluoride (VDF) to the applications of VDF-containing polymers and copolymers: Recent developments and future trends, *Chem. Rev.*, **109**, 6632–6686.

17. Frick, A., Sich, D., Heinrich, G. *et al.* (2012). Classification of new melt-processable PTFE: Comparison of emulsion- and suspension-polymerized materials, *Macromol. Mater. Eng.*, **297**, 329–341.

18. Drobny, J.G. (2009). *Technology of Fluoropolymers, 2ⁿᵈ ed.*, CRC Press, New York, p. 227.

19. deMontigny, D., Tontiwachwuthikul, P. and Chakma A. (2006). Using polypropylene and polytetrafluoroethylene membranes in a membrane contactor for CO_2 absorption, *J. Membr. Sci.*, **277**, 99–107.

20. Nishikawa, N., Ishibashi, M., Ohta, H. *et al.* (1995). CO_2 removal by hollow-fiber gas-liquid contactor, *Energy Convers. Manage.*, **36**, 415–418.

21. Franco, J.A., Kentish, S.E., Perera, J.M. *et al.* (2011). Poly(tetrafluoroethylene) sputtered polypropylene membranes for carbon dioxide separation in membrane gas absorption, *Ind. Eng. Chem. Res.*, **50**, 4011–4020.

22. Goessi, M., Tervoort, T. and Smith, P. (2007). Melt-spun poly(tetrafluoroethylene) fibers, *J. Mater. Sci.*, **42**, 7983–7890.

23. Xiong, J., Huo, P. and Ko, F.K. (2009). Fabrication of ultrafine fibrous polytetrafluoroethylene porous membranes by electrospinning, *J. Mater. Res.*, **24**, 2755–2761.

24. Huang, Q., Xiao, C. and Hu, X. (2010). A novel method to prepare hydrophobic poly(tetrafluoroethylene) membrane, and its properties, *J. Mater. Sci.*, **45**, 6569–6573.

25. Huang, Q., Xiao, C., Hu, X. *et al.* (2011). Study on the effects and properties of hydrophobic poly(tetrafluoroethylene) membrane, *Desalination*, **277**, 187–192.
26. Takagi, Y., Lee, J.C., Yagi, S. *et al.* (2011). Fiber making directly from poly(tetrafluoroethylene) emulsion, *Polymer*, **52**, 4099–4105.
27. Drobny, J.G. (2009). *Technology of fluoropolymers, 2nd edn.*, CRC Press, New York, p. 227.
28. Halar^{®} (2012). ECTFE ethylene-chlorotrifluoroethylene design and processing guide, available at: www.solvaysolexis.com, [accessed Nov. 2012].
29. Simone, S., Figoli, A., Santoro, S. *et al.* (2012). Preparation and characterization of ECTFE solvent resistant membranes and their application in pervaporation of toluene/water mixtures, *Sep. Purif. Technol.*, **90**, 147–161.
30. Ramaswamy, S., Greenberg, A.R. and Krantz, W.B. (2002). Fabrication of poly(ECTFE) membranes via thermally induced phase separation, *J. Membr. Sci.*, **210**, 175–180.
31. Roh, I.J., Ramaswamy, S., Krantz, W.B. *et al.* (2010). Poly(ethylenechlorotrifluoroethylene) membrane formation via thermally induced phase separation (TIPS), *J. Membr. Sci.*, **362**, 211–220.
32. Müller, H.J. (2006). A new solvent resistant membrane based on ECTFE. *Desalination*, **199**, 191–192.
33. Maccone, P. and Fossati, P. (2003). Semipermeable porous membranes of semi-crystalline fluoropolymers. US Patent 6,559,192 B2 (2003).
34. Mutoh Y, Miura M. (1987). Porous fluorine resin membrane and process for preparing the same, US Patent 4,702,836.
35. Mullette, D. and Muller, H.J. (2009). Poly(ethylene chlorotrifluoroethylene) membranes, US Patent 7,632,439 B2.
36. Prabhakar, R.S., Freeman, B.D. and Roman, I. (2004). Gas and vapor sorption and permeation in poly(2,2,4-trifluoro-5-trifluoromethoxy-1,3-dioxole-co-tetrafluoroethylene), *Macromolecules*, **37**, 7688–7697.
37. Jansen, J.C., Friess, K. and Drioli, E. (2011). Organic vapour transport in glassy perfluoropolymer membranes: A simple semi-quantitative approach to analyze clustering phenomena by time lag measurements, *J. Membr. Sci.*, **367**, 141–151.
38. Gugliuzza, A. and Drioli, E. (2007). PVDF and HYFLON AD membranes: Ideal interfaces for contactor applications, *J. Membr. Sci.*, **300**, 51–62.
39. Gugliuzza, A., Ricca, F. and Drioli, E. (2006). Controlled pore size, thickness and surface free energy of super-hydrophobic PVDF^{®} and Hyflon^{®} AD membranes, *Desalination*, **200**, 26–28.
40. Fang, Z.Z. (ed.) (2010). *Sintering of Advanced Materials. Fundamentals and Processes*, Woodhead Publishing Limited, Cambridge, UK.
41. Fink, D. (ed.) (2004). *Transport Processes in Ion-Irradiated Polymers*, Springer, Berlin.
42. Hildebrand, J. and Scott, R. (1949). *Solubility of Nonelectrolytes*, Reinhold, New York.
43. Hansen, C. (1969). The universality of the solubility parameters, *Ind. Eng. Chem. Prod. Res. Dev.*, **8**, 2.
44. Bottino, A., Camera-Roda, G., Capannelli, G. *et al.* The formation of microporous polyvinylidene difluoride membranes by phase separation, *J. Membr. Sci.*, **57**, 1–20.
45. Mulder, M. (2000). *Basic Principles of Membrane Technology*, Kluwer Academic Publishers, Dordrecht, The Netherlands.

46. van Krevelen, D.W. and Hoftyzer, P.J. (1976). *Properties of Polymers: Their Estimation and Correlation with Chemical Structure*, 2nd edn., Elsevier, Amsterdam.
47. Flory, P.J. (1953). *Principles of Polymer Chemistry*, Cornell University Press, Ithaca, NY.
48. Frommer, M.A., Feiner, I., Kedem, O. *et al.* (1970). The mechanism for the formation of 'skinned' membranes. II. Equilibrium properties and osmotic flows determining membrane structure, *Desalination*, **7**, 393.
49. Guillotin, M., Lemoyne, C., Noel, C. *et al.* (1977). Physicochemical processes occurring during the formation of cellulose diacetate membranes. Research of criteria for optimizing membrane performance. IV. Cellulose diacetate-acetone-additive casting solutions, *Desalination*, **21**,165.
50. Strathmann, H. and Kock, K. (1977). The formation mechanism of phase inversion membranes, *Desalination*, **21**, 241.
51. Frommer, M.A., and Messalem, R.M. (1973). Mechanism of membrane formation. VI. Convective flows and large void formation during membrane precipitation, *Ind. Eng. Chem. Prod. Res. Dev.*, **12**, 328.
52. Cohen, C., Tanny, G.B. and Prager, S. (1979). Diffusion-controlled formation of porous structure in ternary polymer systems, *J. Polym. Sci., Polym. Phys. Ed.*, **17**, 477.
53. Friedrich, G., Driancourt, A., Noel, C. *et al.* (1981). Asymmetric reverse osmosis and ultrafiltration membranes prepared from sulfonated polysulphone, *Desalination*, **36**, 39.
54. Wijmans, J.G., Kant, J., Mulder, M.H.V. *et al.* (1985). Phase separation phenomena in solutions of polysulfone mixtures of a solvent and a nonsolvent: Relationship with membrane formation, *Polymer*, **26**, 1539.
55. Hakagawa, K. and Ishida, Y. (1973). Annealing effect in poly(vinylidene fluoride) as revealed from specific volume measurements, differential scanning calorimetry and electron microscopy, *J. Polym. Sci. Polym. Phys. Ed.*, **11**, 2153.
56. Yilmaz, L. and McHugh, A.J. (1986). Modelling of asymmetric membrane formation. I. critique of evaporation models and development of diffusion equation formalism for the quench period, *J. Membr. Sci.*, **28**, 287.
57. Lloyd, D.R., Kinzer, K.E. and Tseng, H.S. (1990). Microporous membrane formation via thermally induced phase separation. I. Solid-liquid phase separation, *J. Membr. Sci.*, **52**, 239–261.
58. Tomaszewska, M. (1996). Preparation and properties of flat-sheet membranes from poly(vinylidene fluoride) for membrane distillation, *Desalination*, **104**, 1–11.
59. Feng, C., Shi, B., Li, G. *et al.* (2004). Preliminary research on microporous membrane from F2.4 for membrane distillation, *Sep. Purif. Technol.*, **39**, 221–228.
60. Feng, C., Shi, B., Li, G. *et al.* (2004). Preparation and properties of microporous membrane from poly(vinylidene fluoride-co-tretrafluoroethylene) (F2.4) for membrane distillation, *J. Membr. Sci.*, **237**, 15–24.
61. Bottino, A., Capannelli, G., Munari, S. *et al.* (1988). High performance ultrafiltration membranes cast from LiCl doped solutions, *Desalination*, **68**, 167.
62. Arcella, V., Colaianna, P., Maccone, P. *et al.* (1999). A study on a perfluoropolymer purification and its application to membrane formation, *J. Membr. Sci.*, **163**, 203–209.
63. Xu, J.B., Lange, S., Bartley, J.P. *et al.* (2004). Alginate-coated microporous PTFE membranes for use in the osmotic distillation of oily feeds, *J. Membr. Sci.*, **240**, 81–89.

64. Wu, Y., Kong, Y., Lin, X. *et al.* (1992). Surface-modified hydrophilic membranes in membrane distillation, *J. Membr. Sci.*, **72**, 189–196.
65. Favia, P. and D'Agostino, R. (1998). Plasma treatments and plasma deposition of polymers for biomedical applications, *Surface and Coatings Tech.*, **98**, 1102–1106.
66. Kong, Y., Lin, X., Wu, Y. *et al.* (1992). Plasma polymerisation of octafluorocyclobutane and hydrophobic microporous composite membranes for membrane distillation, *J. Applied Polymer Sci.*, **46**, 191–199.
67. Khayet, M., Mengual, J.I. and Matsuura, T. (2005). Porous hydrophobic/hydrophilic composite membranes. Application in desalination using direct contact membrane distillation, *J. Membr. Sci.*, **252**, 101–113.
68. Khayet, M., Suk, D.E., Narbaitz, R.M. *et al.* (2003). Study on surface modification by surface-modifying macromolecules and its applications in membrane-separation processes, *J. Appl. Polym. Sci.*, **89**, 2902–2916.
69. Li, D. and Newmann, A.W. (1992). Equation of state for interfacial tensions of solid-liquid systems, *Advances Coll. Interf. Sci.*, **39**, 299–345.
70. Franken, A.C.M., Nolten, J.A.M., Mulder, M.H.V. *et al.* (1987). Wetting criteria for the applicability of membrane distillation, *J. Membr. Sci.*, **33**, 315–328.
71. Courel, M., Tronel-Peyroz, E., Rios, G.M. *et al.* (2001). The problem of membrane characterization for the process of osmotic distillation, *Desalination*, **140**, 15–25.
72. Troger, J., Lunkwitz, K. and Burger, W. (1997). Determination of the surface tension of microporous membranes using contact angle measurements, *J. Coll. Interf. Sci.*, **194**, 281–286.
73. Good, R.J. and van Oss, C.J. (1992). 'Modern approach to wettability', in Schrader, M.E. and Loeb, G.L. (eds.), *The Modern Theory Contact Angle and the Hydrogen Bond Components of Surface Energies*, Plenum Press, New York.
74. De Bartolo, L., Morelli, S., Bader, A. *et al.* (2002). Evaluation of cell behaviour related to physico-chemical properties of polymeric membranes to be used in bioartificial organs, *Biomaterials*, **23**, 2485–2497.
75. Tomaszewska, M. (1999). Membrane distillation, *Envir. Protect. Eng.*, **25(1–2)**, 37–47.
76. De Bartolo, L., Morelli, S., Rende, M. *et al.* (2004). New modified polyetheretherketone membrane for liver cell culture in biohybrid systems: Adhesion and specific functions of isolated hepatocytes, *Biomaterials*, **25**, 3621–3629.
77. Lawson, K.W. and Lloyd, D.R. (1997). Membrane distillation, *J. Membr. Sci.*, **124**, 1–25.
78. Reed, B.W., Semmens, M.J. and Cussler, E.L. (1995). 'Membrane contactors', in Noble, R.D. and Stern, S.A. (eds.), *Membrane Separation Technology. Principles and Applications*, Elsevier, Amsterdam.
79. Schneider, K., Holz, W. and Wollbeck, R. (1988). Membranes and modules for trans-membrane distillation, *J. Membr. Sci.*, **39**, 25–42.
80. Pena, L., Ortiz de Zarate, J.M. and Mengual, J.I. (1993). Steady states in membrane distillation: influence of membrane wetting, *J. Chem. Soc. Faraday Trans.*, **89**, 4333–4338.
81. Gostoli, C. and Sarti, G.C. (1989). Separation of liquid mixtures by membrane distillation, *J. Membr. Sci.*, **41**, 211–224.
82. Schofield, R.W., Fane, A.G. and Fell, C.J.D. (1990). Gas and vapour transport through microporous membranes. I: Knudsen-Poiseuille transition, *J. Membr. Sci.*, **53**, 159–171.

83. Martinez-Diez, L. and Vazquez-Gonzalez, M.I. (1996). Temperature polarization in mass transport through microporous membranes, *AIChE J.*, **42**, 1844–1852.
84. Schneider, K., Holz, W. and Wollbeck, R. (1988). Membranes and modules for transmembrane distillation, *J. Membr. Sci.*, **39**, 25–42.
85. Laganà, F., Barbieri, G. and Drioli, E. (2000). Direct contact membrane distillation: Modelling and concentration experiments, *J. Membr. Sci.*, **166**, 1–11.
86. Phattaranawik, J., Jiraratananon, R. and Fane, A.G. (2003). Effect of pore-size distribution and air flux on mass transport in direct contact membrane distillation, *J. Membr. Sci.*, **215**, 75–85.

Chapter 4

Membrane Crystallization
of Inorganic Compounds

4.1 Introduction

Although crystallizers have been in operation for many decades in the chemical industry, their importance is still relevant due to the large variety of materials that are marketed in crystalline form. Moreover, crystallization is a practical method for obtaining pure chemical substances from relatively impure solutions in a solid form.

If crystals need to be further processed in reaction with other compounds or simply filtered from their mother liquor, washed, stored, etc., then, a uniform size is required. Even if crystals are commercialized directly as a final product, market standards require individual crystals to be non-aggregated, non-caked, and uniform in size and shape. Therefore, crystal size distribution (CSD) is a crucial parameter to be carefully controlled when designing or choosing a crystallizer.

Crystals, as a result of a specific and characteristic three-dimensional arrangement of molecules, ions, or atoms, often appear as polyhedrons having well-defined geometry with flat faces and sharp angles. The shape of a crystals is not merely an aesthetic question; it severely impacts the packaging and the filterability of the solid particles, since crystals with higher specific surface area tend to easily adhere each other due to the higher amount of mother liquor retained.

In this respect, the design and operation of large-scale industrial crystallizers still pose many problems. Large-scale plants often produce crystals that do not satisfy the requested quality criteria — more and more stringent due to the shift of the market trend from base chemicals towards life-science products. Critical issues in the design of an industrial crystallization process come from mutual interactions between complex kinetic aspects (rate of supersaturation generation, nucleation, and growth) and fluid dynamics. On a large scale, process parameters impacting nucleation rate, such as temperature, supersaturation, and velocity profile, are typically not uniformly distributed in the crystallizer body. As a consequence, the slurry is not well-mixed and suspended, with negative impact on the quality of the final product.[1]

At industrial level, evaporative crystallization from solution is substantially dominated by circulating magma crystallizers in two basic variants: the forced circulation (FC) crystallizer and the Draft-Tube-Baffled (DTB) crystallizer.

In an FC crystallizer, the externally heated suspension is pumped to the boiling zone through either a tangential or an axial inlet. Due to the high recirculation rate imposed, with the aim of limiting segregation of the supersaturated solution and to keep crystal in suspension throughout the crystallizing zone, the main design problem is the formation of vortices; this leads to thermal short-circuiting that results in different levels of supersaturation within the body, variable nucleation rates, and consequent cycling of the CSD between fine and coarse particles as the crystal surface varies.[2,3]

In a DTB crystallizer, the suspension of crystals is maintained by a large, slow-moving propeller surrounded by draft tubes or baffles within the body. In this apparatus, the final shape of the crystals is often unsatisfactory, as a consequence of the fact that larger crystals are rounded off by abrasion and attrition with the mechanical parts of the apparatus. In most crystallizers a vacuum is used in order to promote adiabatic evaporative cooling and to generate supersaturation.

In a membrane crystallizer, microporous hydrophobic membranes are used to generate and control the evaporative mass transfer of volatile solvents in order to concentrate solutions above the saturation limit, thus attaining a homogeneously supersaturated environment throughout the whole hollow-fiber bundle.[4]

In a continuous apparatus (Fig. 4.1), slurry leaving the crystallizer body is mixed to the feed and pumped through a heat exchanger that increases its temperature typically to 40°C–50°C. The heated mother liquor is recirculated through the lumen side of the membrane module; here the solvent (typically water) partially evaporates, diffuses across the membrane, and condenses on the opposite side. Heterogeneous nucleation takes place at the membrane surface; once nuclei are formed, they are gently detached due to the shear stress generated by the recirculating flow, and then grow on a separate tank (crystallizer body — no evaporation here). Slurry flow rate, the size of the body, and temperature profiles are crucial parameters, which must be carefully set in order to prevent membrane blocking due to crystal deposition.

In a membrane crystallizer, the laminar flow of the mother liquor through the fibers of the modules results in a low shear stress and in an improved homogeneity

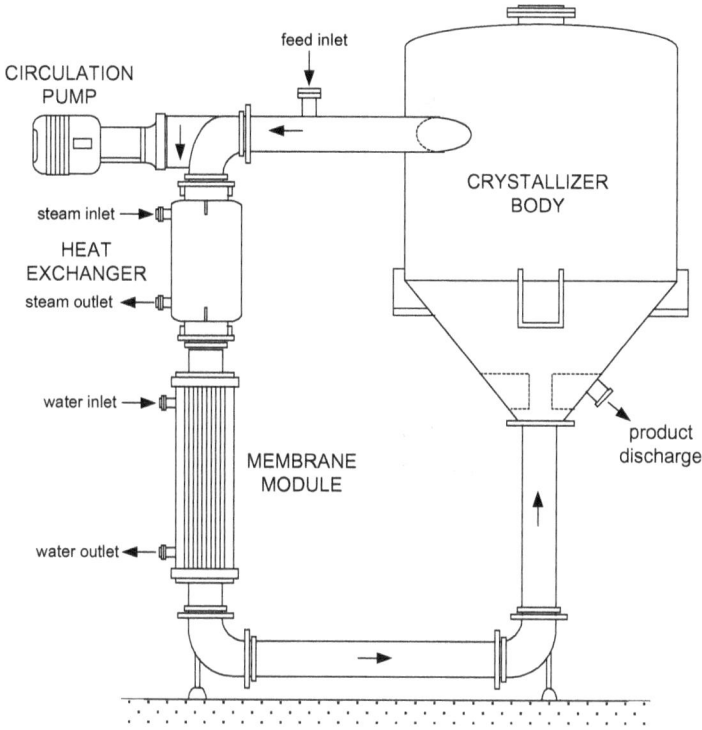

Figure 4.1. Membrane crystallizer. (Reprinted from [5] with permission. Copyright © 2005 Elsevier B.V.).

of the solution; this is expected to promote a well-ordered organization of the molecules, thus resulting in the formation of crystals with regular structural properties and sharp CSD. A comparison among cubic sodium chloride crystals produced by different crystallizers is reported in Fig. 4.2.

4.2 Product Characterization: Shape and CSD

CSD is a critical parameter for industrial crystallization, also determining how easy and efficient is the preliminary separation of solids from the liquid slurry, and the subsequent drying. Relatively large crystals limit the adhesion of mother liquor after filtration and, therefore, show a low tendency toward caking. On the contrary, small crystals are preferentially required whenever reduced dissolution times are needed. In all cases, narrow distributions are necessary (especially in pharmaceutical applications).

| (a) | (b) | (c) |

Figure 4.2. Typical NaCl crystals produced by (a) a FC crystallizer, (b) a draft-tube-baffle crystallizer; (c) a membrane crystallizer.

A preliminary difficulty in determining CSD is the definition of the crystal size due to the fact that crystals may exhibit a large variety of shapes (for example, cubic, spherical, platelet or needle-like). It is therefore recommended to define a characteristic length of the crystals to be related to crystal surface area or volume using appropriate shape factors.

Assuming that L_C is the characteristic dimension (length) of a crystal having a total surface area A_C and a total volume V_C, we have

$$\alpha = \frac{V_C}{L_C^3} \tag{4.1a}$$

and

$$\beta = \frac{A_C}{L_C^2}, \tag{4.1b}$$

where α and β are the volume shape factor and the surface shape factor, respectively. The shape factor is clearly correlated to the mass of a single crystal (M_c):

$$M_C = \alpha \cdot \rho_C \cdot L_C^3, \tag{4.2}$$

where ρ_C is the crystal density. In Table 4.1 a few examples of shape factors for the few simplest crystals geometries are reported.

The most common way to illustrate the CSD of a solid product is to plot the relative percentage frequency curve $f(L_i)$ vs the size classes L_i. The function $f(L_i)$ is defined as the ratio between the "counted crystals" belonging to a certain size class (N_i) and the total number of crystals measured ($\sum_i N_i$) divided by the width of the size class (ΔL_i):

$$f(L_i) = \frac{N_i}{\sum_i N_i} \cdot \frac{1}{\Delta L_i}. \tag{4.3}$$

Table 4.1.   Examples of crystals shape factors.

Geometric shape	α	β	Example of crystalline habit
Cube	1.000	6.000	Sodium chloride crystals[†]
Sphere	0.524	3.142	Calcium carbonate ($CaCO_3$) crystal[†]
Tetrahedron	0.118	1.732	Hen egg white lysozyme[†]
Hexagonal prism	0.867	5.384	Magnesium sulphate heptahydrate
Platelet ($10 \times 1 \times 1$)	0.100	2.400	Porcine pancreatic trypsin[†]

(Continued)

Table 4.1.   (*Continued*)

Geometric shape	α	β	Example of crystalline habit
Needle-like ($10 \times 1 \times 1$)	0.010	0.420	Calcium sulfate crystals[†,‡]

[†]Grown on microporous polypropylene membrane.
[‡]Grown on microporous polyvinylidene difluoride membrane.

The advantage in using relative percentage is that values are independent from the number of crystals counted. The most frequently occurring size, represented by the abscissa value in correspondence of the maximum in the $q(L_i)$ curve, is the "mode" of the distribution. An example of a frequency distribution curve is given in Fig. 4.3 for sodium chloride crystals obtained in a membrane crystallization (MCr) experiment carried out on a seawater reverse osmosis (SWRO) brine.

CSD can be also conveniently represented by a cumulative distribution function (F) defined as

$$F = \int_{L_{min}}^{L} f(L_i)dL = \sum_{L_i=L_{min}}^{L} f(L_i)\Delta L_i. \tag{4.4}$$

F represents the percentage of the measured crystals that are smaller than a certain size. Figure 4.4 shows the cumulative size distribution function for sodium sulfate crystals (in the form of thenardite) grown in a membrane crystallizer. The value of the abscissa in corresponding to $F = 50\%$ gives the median size of the CSD.

The crystal length distribution at 30 min (Fig. 4.4), spanning between 45 μm and 165 μm, is comparable with data reported by Klug *et al.* (1989) in a flow-cell batch crystallization operated at higher supersaturation ratio (0.088) and temperature (61°C).[8] This result proves that the accelerated crystallization kinetics observed in the membrane crystallizer is an effect of the heterogeneous nucleation occurring at the polymeric surface of the membrane.

The mass distribution $W(L_i)$, expressed in terms of the mass fraction of crystals belonging to a certain size class divided by the width of the size class (ΔL_i) is

$$W(L_i) = \alpha \rho_c L_i^3 \frac{\Delta N_i}{\Delta L_i}, \tag{4.5}$$

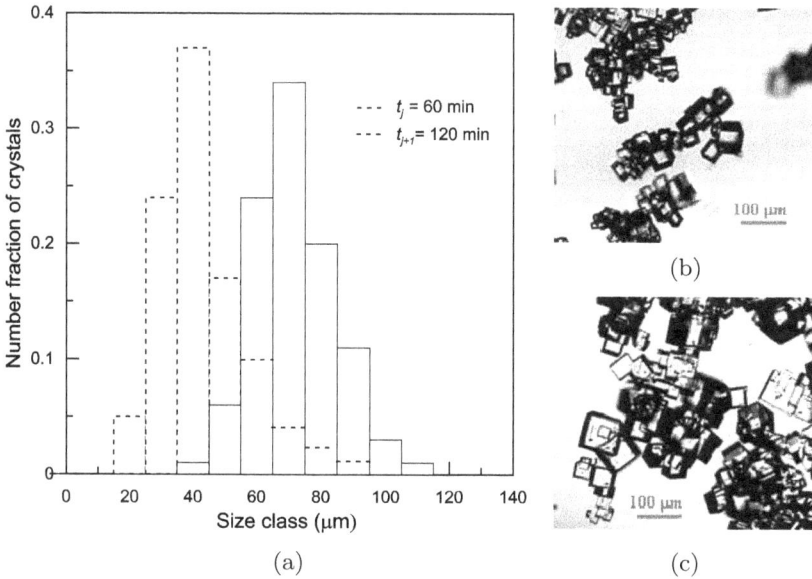

Figure 4.3. (a) Size distribution of NaCl crystals grown from natural reverse osmosis (RO) brines. Images of crystals at (b) 60 min, (c) 120 min after reaching supersaturation. (Reprinted from [6] with permission. Copyright © 2009 Elsevier B.V.)

where α is the volume shape factor, ρ_c is the crystal density, and ΔN is the number of crystals — counted in a unit volume of slurry — having length between L and $L + \Delta L$.

Apart from the average sizes evaluated by distribution diagrams, the width of a distribution around the mean size is often characterized by the coefficient of variation (CV), equal to the standard deviation divided by the mean size:

$$CV = \left\{ \left[\frac{\int_0^\infty (L - L_{50\%})^2 W(L_i)dL}{\int_0^\infty W(L_i)dL} \right]^{\frac{1}{2}} \right\} \cdot \frac{\int_0^\infty W(L_i)dL}{\int_0^\infty LW(L_i)dL}, \qquad (4.6)$$

where $L_{50\%}$ is the crystal length corresponding to a value of $W(L)$ equal to 50%. A narrow CSD is characterized by a low CV.

Under the assumption that CSD distribution can be approximated by a Gaussian function, Eq. (4.6) can be approximated as

$$CV = \frac{L_{16\%} - L_{84\%}}{2 \cdot L_{50\%}} \cdot 100, \qquad (4.7)$$

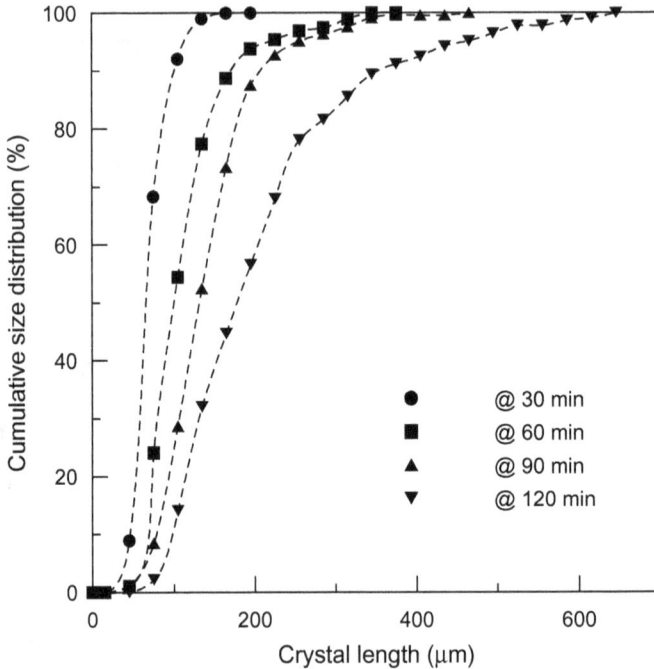

Figure 4.4. Cumulative size distribution of thenardite $Na_2SO_4(V)$. (Reprinted from [7] with permission. Copyright © 2010 Elsevier B.V.)

where the $L_{16\%}$, $L_{50\%}$, and $L_{84\%}$ values can be taken from the plot of the cumulative function F.

Experimental CV values measured in tests carried out on NaCl range from 15%–30%.[9] For Na_2SO_4 crystallization, crystal samples collected at 30 minutes showed the lowest CV (23%).[8] In general, the dispersion of CSD increases with time and reaches a constant value of 40% ± 1% after 90 minutes (typical residence time in industrial crystallizers). As a comparison, the coefficient of variation for ideal mixed-suspension, mixed-product removal (MSMPR) is 50% for size-independent growth, but becomes significantly higher for size-dependent growth.[10]

4.3 Kinetic Aspects: The Mixed-Suspension Mixed-Product Removal (MSMPR) Model

The kinetic parameters of industrial crystallization devices (including effective nucleation rate B_0 and crystal growth rate G) are generally determined by taking an ideally continuous MSMPR crystallizer as a reference model.

The adoption of an MSMPR model imposes some restrictions in terms of operative conditions. The most significant requisite is that the slurry (mother solution and crystals in suspension) is perfectly mixed; therefore, in principle, no differences in supersaturation, temperature, magma density, and CSD are supposed to exist within the body of the crystallizer. Moreover, it is assumed that crystals are removed isokinetically from the perfectly mixed MSMPR without any classification of the suspension. The crystallizer operates under steady-state conditions: no volume variation occurs and processes such as agglomeration, attrition, breakage, dissolution, etc. are absent.

The population balance equation (i.e. the number of crystals per particle size per volume of solution, n), if the crystal growth rate is size-independent, can be written as[11]

$$G \cdot \frac{dn(L)}{dL} + \frac{n(L)}{\tau} = 0, \tag{4.8}$$

to be solved under the following boundary conditions:

$$n(L, t = 0) = 0 \tag{4.9a}$$

$$n(L = 0, t) = \frac{B_0}{G} = n_0. \tag{4.9b}$$

Integration of Eqs. (4.8)–(4.9a–4.9b) provides a quite simple mathematical relationship between the population density n, the crystal size L, crystal growth rate G, and residence time of solids in the crystallizer τ:

$$\frac{n(L)}{n_0} = \exp\left(-\frac{L}{G \cdot \tau}\right), \tag{4.10a}$$

or, if a linear plot is preferred:

$$\ln\left[\frac{n(L)}{n_0}\right] = -\frac{L}{G \cdot \tau}. \tag{4.10b}$$

Therefore, the growth rate G can be deduced from the slope of the straight line $\ln(n)$ as a function of L, which is equal to $-1/(G \cdot \tau)$, while the intercept at $n = 0$ gives the value of $\ln(n_0)$ and consequently, from Eq. (4.9b), the nucleation rate $B_0 = n_0/G$.

A cumulative mass distribution of crystals Q_m can be derived from the population balance equation of an ideal MSMPR crystallizer:

$$Q_m = 1 - \frac{1 + \xi_i + \left(\frac{\xi_i^2}{2}\right) + \left(\frac{\xi_i^3}{3}\right)}{\exp(\xi_i)}, \tag{4.11}$$

where $\xi_i = L_i/(G \cdot \tau)$.

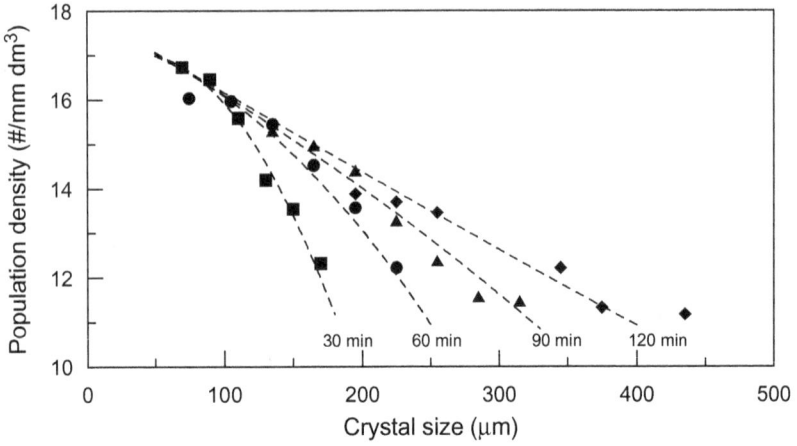

Figure 4.5. Population density vs. crystal size as measured within 2 hours for Na$_2$SO$_4$ membrane crystallization (Reprinted from [7] with permission. Copyright © 2010 Elsevier B.V.).

The population density distribution of Na$_2$SO$_4$ crystals produced in membrane crystallization batch tests is reported in Fig. 4.5 at 4 time intervals between 30 minutes and 2 h of operation. Tun *et al.* (2005), who first carried out membrane distillation crystallization tests on sodium sulfate, reported a value of about 10 #/mm · cm^3 (corresponding to ∼16.9 #/mm · dm^3) at 45 minutes and crystal length of 100 μm, which is in good agreement with the value of 16.5 #/mm · dm^3 found in our study at 30 minutes and the same crystal length.[12] The almost linear shape of population density curves confirms that the behavior of membrane crystallization system approaches — with good approximation — the MSMPR model.

The moments of the population density distribution are a source of additional information. The 0th order moment m_0 gives the total number of crystals per unit of volume suspension N_{tot}:

$$m_0 = \int_0^\infty n(L)dL \qquad (4.12a)$$

and $N_{tot} = n_0 \cdot G \cdot \tau = B_0 \cdot \tau$. The 2nd order moment m_2 gives the total surface area per unit volume of suspension $a_{tot} = \beta \cdot m_2$:

$$m_2 = \int_0^\infty L^2 n(L)dL \qquad (4.12b)$$

and $a_{tot} = 2\beta n_0 (G \cdot \tau)^3$. The 3rd order moment m_3 gives the volume of crystals per unit volume of suspension $\varphi = \alpha \cdot m_3$, which is conveniently expressed in terms of suspension density through the crystal density ρ_c as $m_T = \rho_c \cdot \varphi$:

$$m_3 = \int_0^\infty L^3 n(L)dL \qquad (4.12c)$$

and $m_T = 6\alpha \cdot \rho_c \cdot n_0 (G \cdot \tau)^4$.

4.4 Growth Rate

The overall growth rate G is usually related to the overall concentration driving force Δc by an empirical relation:

$$G = k_g \Delta C^g, \qquad (4.13)$$

where g is the overall order of the growth process and k_g is the growth rate constant that is a function of temperature, controlling the kinetic mechanism (diffusion or integration) and presence of impurity in the system (Fig. 4.6).

The process of crystal growth occurs according to sequential stages: transport of molecule from the bulk of the mother liquor to the crystal surface; adsorption and

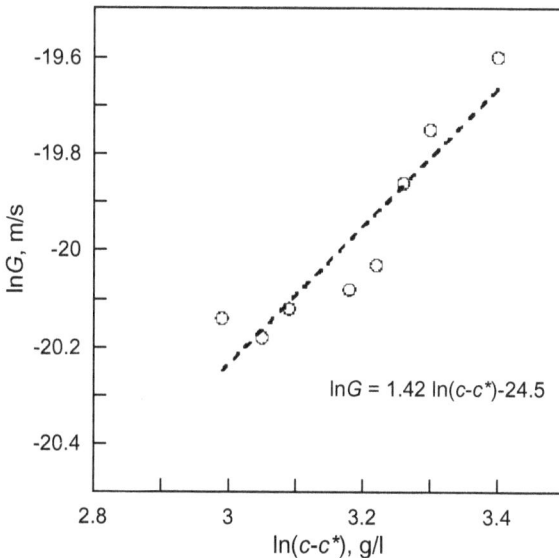

Figure 4.6.   Correlation between growth rate and supersaturation (c = actual concentration, $c*$ = saturation) for NaCl in a membrane crystallizer.

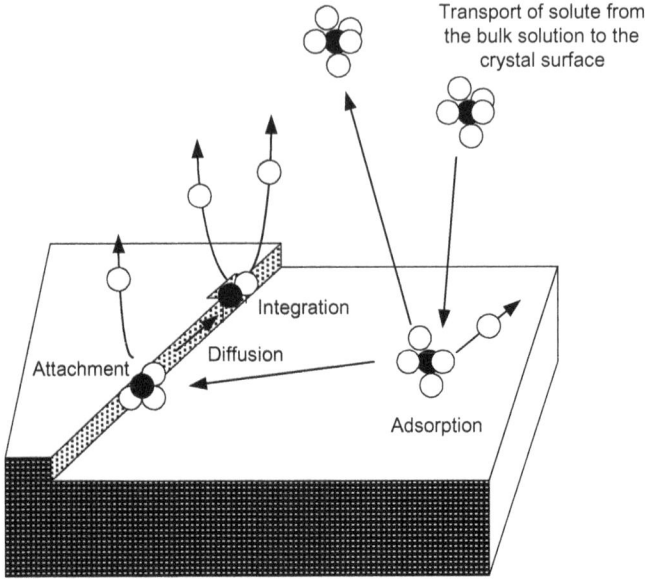

Figure 4.7. Exemplification of sequential stages of crystal growth.

diffusion over the surface; attachment and diffusion along a step; and integration into the crystal at a kink site (Fig. 4.7).

When diffusive phenomena in solution are the rate-determining step for crystal growth, a concentration gradient between the bulk and the crystal surface is established; in this case, Fick's first and second laws can be used to describe the diffusion of a molecule through the boundary layer to a stationary particle. In a membrane crystallizer under forced flow, diffusion is enhanced by convective mass transport, and the particle flux J toward a crystal growing in a supersaturated environment is generally expressed as

$$J = k_x(c_\infty - c_e) = \frac{D}{\delta}(c_\infty - c_e), \qquad (4.14)$$

where k_x is the mass transfer coefficient, D is the diffusion coefficient, δ is the boundary layer thickness, and c_∞ and c_e are the bulk and surface (equilibrium) concentrations, respectively. This yields a convective growth rate G given by

$$G = \xi^{-1}k_x c_e(S - 1), \qquad (4.15)$$

where ξ is the number of molecules precipitated per unit volume of solution and S is the supersaturation ratio ($S = c_\infty/c_e$).

The boundary layer thickness δ is evaluated by using Carlson analysis of laminar flow with a slip velocity u:[5]

$$\delta = \left(\frac{L}{0.282}\right) \text{Re}^{-1/2} Sc^{-1/3} \left(\frac{L}{0.232}\right) \left(\frac{uL}{v}\right)^{-1/2} \left(\frac{v}{D}\right)^{-1/3}, \tag{4.16}$$

where L can be approximated by the crystal linear dimension and v is the kinematic viscosity. The most interesting feature of this theory is that δ is predicted to be inversely proportional to the square root of the velocity and that G should increase proportionally to $u^{1/2}$.

The Pèclet number is the control parameter for the dominant mechanism, defined as the ratio of rate constants for nonlinear integration kinetics and bulk transport in the form of

$$Pe_k = \beta_f \frac{\delta}{D}, \tag{4.17}$$

where β_f is the face kinetic coefficient. Decreasing values of Pe_k correspond to kinetic controlled regimes with significantly faster transport. Under kinetic control, the incorporation of impurities is a potential source of defects and aggregates, resulting in growth rate reduction and quality degradation. According to these considerations, a deceleration of crystal growth due to an enhanced supply of impurities to the interface is expected.

Under the assumption that impurities adsorbed at the crystal surface prevent the step advancement during layer growth, the Cabrera–Vermilyea model[13] modified by Kubota and Mullin[14] was used to correlate the relative growth rate (G/G_0) to the concentration of humic acid (HA) (c):

$$\frac{G}{G_0} = 1 - \alpha \frac{Kc}{1 + Kc}, \tag{4.18}$$

where α is an effectiveness factor ($\alpha \to 1$ if the relative growth rate approaches zero for increasing coverage) and K is the Langmuir constant. The calculated K was 0.09 l/mg ($\alpha = 0.87$) for vaterite crystals grown in the presence of a polymeric membrane. This value, if compared to data reported by Westin *et al.* (2005), is higher than those found for citric acid, diethylenetriaminepentaacetic acid (DTPA) and ethylenediaminetetraacetic acid (EDTA) (0.07 l/mg, 0.019 l/mg, and 0.027 l/mg, respectively),[15] suggesting that $CaCO_3$ growth is affected by surface adsorption of HA and, to a major extent, when crystal nucleation and growth is mediated by a membrane surface.

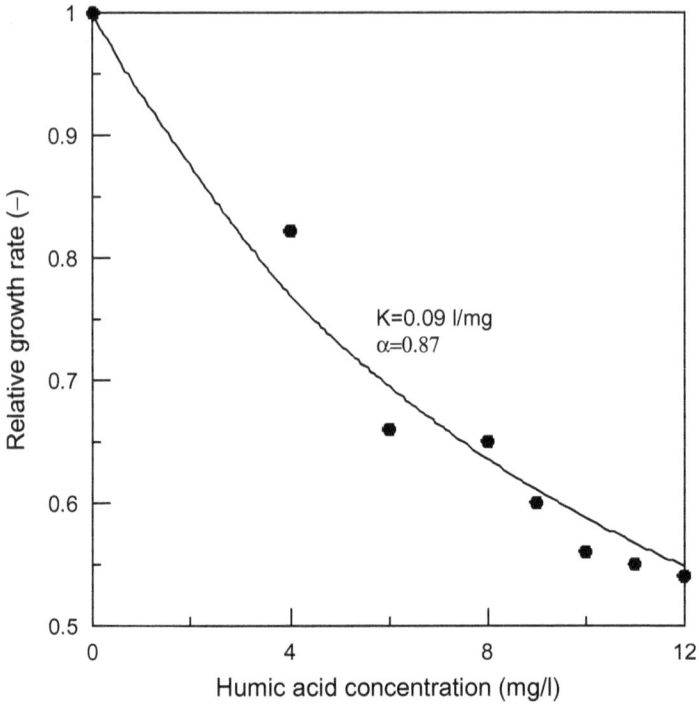

Figure 4.8. Relative growth rate of $CaCO_3$ as a function of HA concentration in presence of the membrane. (Reprinted from [16] with permission. Copyright © 2009 Elsevier B.V.)

4.5 Applications

4.5.1 *Treatment of brines*

The potential of MDCr technology for simultaneous production of pure water and salt crystals from brine solution has recently attracted the interest of membranologists. In a pioneering study, Curcio *et al.* (2001) investigated the crystallization of NaCl solutions using microporous polypropylene membranes; the operative temperatures were 29°C at the feed inlet and 9°C at the distillate inlet. It was observed that, from an initial value of approximately 0.5 l/m²h, the transmembrane flux progressively decreased in concomitance with the increase of NaCl concentration in the retentate and the consequent reduction of the activity coefficient. At the end of the test, a flux reduction of 20% was observed. The maximum achieved magma density was 3.8 kg of NaCl per m³ of mother liquor; CSD exhibited a CV of 42.3%.[4]

MCr operations carried out at high feed temperature, with the aim of increasing the transmembrane flux, resulted in a drastic decrease of performance due to encrustation phenomena observed at the membrane surface. MCr operated between 40°C and 70°C was investigated by Edwie and Chung (2013). Operations carried out at high temperature (60°C–70°C) resulted in a high initial transmembrane flux (close to 9 kg/m²h) gradually decreasing in time (−45% after 200 minutes) due to the occurrences of membrane scaling and wetting facilitated by salt oversaturation at the boundary layer. Stable transmembrane fluxes were observed at temperatures below 50°C. The crystallization kinetics were significantly affected by the feed temperature. The highest yield (34 kg NaCl crystals per 1 m³ mother liquor) was achieved for the maximum investigated feed temperature of 70°C after 200 min operation; the measured nucleation rate was 3.54×10^{10} nuclei/m³·h and the crystal size ranged from 16.37 μm to 131.64 μm. At 40°C the crystal yield decreased to 7.5 kg NaCl/m³, and bigger crystals were obtained (size within the range of 31.31 μm–177.32 μm) concomitantly to a significantly lower nucleation rate (2.21×10^{9} nuclei/m³·h).[17] Figure 4.9 summarizes the trend of NaCl crystal yield and water vapor flux at 40°C and 70°C.

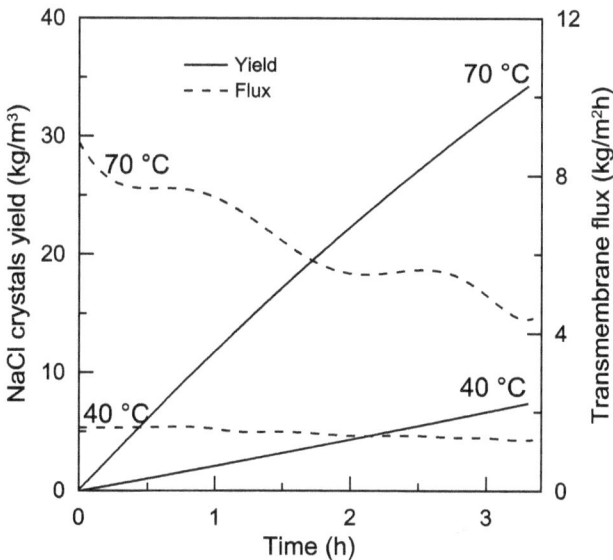

Figure 4.9. Transient trend of NaCl crystals yield and transmembrane flux at feed temperature of 40°C and 70°C (data from [17]).

Table 4.2. SWRO concentrate composition (data from[18]).

Ion	Composition (ppm)[†]
Chloride	45,500
Sodium	25,505
Sulfate	6,507
Magnesium	3,101
Calcium	1,000
Bicarbonate	306
Strontium	26

[†]Seawater intake with salinity of 4.0% and total permeate recovery factor of 51%.

The treatment of brines discharged from SWRO desalination plants has emerged as a potential opportunity for membrane distillation crystallization (MDCr) technology. Increased water recovery factor, reduced volume of concentrates, and the possibility to recover dissolved salts are items that fit well the strategy of zero liquid discharge. A typical composition of seawater concentrated by SWRO plants is reported in Table 4.2.

A thermodynamic analysis of multi-component solutions is generally carried out to estimate which salts can crystallize from brine and in what order. The most critical aspect is related to the reliable estimation of the activity coefficients of ions; this is usually made by Pitzer equations, which today represent the most accurate and reliable calculation method, with evidence of applicability to electrolyte solutions with ionic strength of 6 or higher.[19]

Although the Stiff and Davis saturation index (S&DSI) is the most popular index in industrial desalination, its accuracy is questioned by researchers because the S&DSI it is not applicable to brines characterized by an ionic strength higher than 1. In this respect, saturation index (SI) is claimed to be more reliable.[20]

For a salt $X + Y = XY$,

$$SI = \log\left(\frac{[X][Y]}{K_{sp}}\right) = \log(IAP) - \log(K_{ps}), \tag{4.19}$$

where K_{ps} is the solubility product and IAP is the ion activity product. Referring, for example, to $CaCO_3$ precipitation:

$$K_{ps} = \gamma_{Ca^{2+}}\left[Ca^{2+}\right]\gamma_{CO_3^{2-}}\left[CO_3^{2-}\right] \tag{4.20}$$

$$K_{a2} = \frac{\gamma_{H^+}\left[H^+\right]\gamma_{CO_3^{2-}}\left[CO_3^{2-}\right]}{\gamma_{HCO_3^-}\left[HCO_3^-\right]}. \tag{4.21}$$

Replacing these equations in the definition of SI and rearranging gives

$$SI = \log\left[Ca^{2+}\right] + \log\left[\frac{K_{a2}\left[HCO_3^-\right]}{10^{-pH}}\right] - \log K_{ps} + \log \gamma_{Ca^{2+}} + \log \gamma_{HCO_3^-}.$$

$$(4.22)$$

For a given compound, a positive value of SI indicates that the solution is super-saturated and precipitation of a solid phase may occur; if SI is negative, then the solution is unsaturated with respect to the salt, hence the salt remains dissolved in the solution.

Computer algorithms specifically devoted to addressing the problem of speciation are commercially available or released as freeware; one of the most frequently utilized is PHREEQC, geological solubility software that can be used for highly concentrated solutions. For the brine composition reported in Table 4.2, and a concentration factor of 1, polymorphs of calcium carbonates and calcium sulfates have the highest SI (Table 4.3).

The recovery of dissolved salts from seawater requires an appropriate softening process. Macedonio et al. (2013)[18] treated RO brine (composition in Table 4.2) with sodium carbonate in order to precipitate calcium, sulfate, and carbonate ions. Previous work showed that, when mixing RO brine with Na_2CO_3 with molar ratio of 1:1.05 Ca^{2+}/CO_3^{2-}, 98% of calcium can be removed as $CaCO_3$; the precipitated $CaCO_3$ incorporates a small amount of Mg^{2+} depending on the $[Mg^{2+}]/[Ca^{2+}]$ ratio.[21]

Table 4.3. Saturation index for various species as predicted by PHREEQC. Input: RO brine composition from Table 4.2; concentration factor of 1 (data from[18]).

Compound	SI	Compound	SI
Dolomite – $CaMg(CO_3)_2$	3.91	Hexahydrite – $MgSO_4 \cdot 6H_2O$	−2.57
Magnesite – $MgCO_3$	1.54	Pentahydrite – $MgSO_4 \cdot 5H_2O$	−2.75
Calcite – $CaCO_3$	1.37	Gaylussite – $CaNa_2(CO_3)_2 \cdot 5H_2O$	−2.83
Aragonite – $CaCO_3$	1.20	Pirssonite – $Na_2Ca(CO_3)_2 \cdot 2H_2O$	−2.95
Celestite – $SrSO_4$	0.00	Leonhardite – $MgSO_4 \cdot 4H_2O$	−3.13
Gypsum – $CaSO_4 \cdot 2H_2O$	−0.16	Kieserite – $MgSO_4 \cdot H_2O$	−3.83
Anhydrite – $CaSO_4$	−0.21	Bloedite – $Na_2Mg(SO_4)_2 \cdot 4H_2O$	−4.30
Nesquehonite – $MgCO_3 \cdot 3H_2O$	−1.23	Labile_S – $Na_4Ca(SO_4)_3 \cdot 2H_2O$	−4.34
Halite – $NaCl$	−1.72	Natron – $Na_2CO_3 \cdot 10H_2O$	−4.42
Brucite – $Mg(OH)_2$	−1.94	Bischofite – $MgCl_2 \cdot 6H_2O$	−6.12
Glauberite – $Na_2Ca(SO_4)_2$	−2.09	Trona – $Na_3H(CO_3)_2 \cdot 2H_2O$	−6.68
Mirabilite – $Na_2SO_4 \cdot 10H_2O$	−2.12	Portlandite – $Ca(OH)_2$	−8.26
Nahcolite – $NaHCO_3$	−2.24	Burkeite – $Na_6CO_3(SO_4)_2$	−9.52
Epsomite – $MgSO_4 \cdot 7H_2O$	−2.34		

For the softened RO brine, PHREEQC calculations show that $MgCO_3$ and $SrSO_4$ have the highest SI; however, due to the low Sr^{2+} and CO_3^{2-} concentration in the solution even at high concentration factor, the yield of precipitated $MgCO_3$ and $SrSO_4$ solids is negligible. In increasing recovery factor, the order of precipitated salts is

(1) Halite NaCl — precipitation begins at a concentration factor of 4.4;
(2) Glauberite $Na_2Ca(SO_4)_2$ — precipitation begins at a concentration factor of 7.3;
(3) Brucite $Mg(OH)_2$ — precipitation begins at a concentration factor of 10.99.

The possibility of redesigning important production cycles by combining different membrane operations, available as separation and conversion units, is recognized as a reliable and attractive opportunity due to the synergic effects that can be reached. In this respect, membrane contactor technology today offers additional options to conventional membrane separation units, such as RO, micro-, ultra-, and nano-filtration, etc.

In this framework, the most interesting perspective for implementing membrane distillation/crystallization at industrial scale is its hybridization with other conventional pressure-driven membrane processes in seawater desalination plants. The goal is to increase the water recovery factor and to extract dissolved salts from brines. In fact, seawater is the most abundant aqueous solution on Earth: 3.3% of its composition is represented by dissolved salts, and seven elements (Na, Mg, Ca, K, Cl, S, and Br) account for 93.5% of the ionic species.

The removal of scale components from seawater by nanofiltration (NF) operated upstream RO trains is progressively increasing in reliability. NF membranes are under investigation as pre-treatment systems in order to reduce hardness and total dissolved solids (TDS), and to decrease turbidity: as a result, water recovery factors of about 70% can be achieved. Table 4.4 reports the chemical composition of seawater NF retentate.[22]

Drioli et al. (2004) developed an integrated membrane desalination process able to extract sodium chloride and Epsom salts from NF brine, according to the scheme illustrated in Fig. 4.8. The preliminary objective was to limit the precipitation of calcium carbonate (responsible for scaling) and of calcium sulfate (causing the depletion of SO_4^{2-} ions from solution and, consequently, limiting the recovery of magnesium sulfate). Therefore, Ca^{2+} ions have been precipitated as carbonates by reaction with $NaHCO_3/Na_2CO_3$ solutions. Precipitation of calcium carbonate from artificial NF retentate resulted in the formation of vaterite at pH < 9; calcite was the only solid phase precipitating at higher pH (Fig. 4.9). The formation

Table 4.4. Chemical composition and physical properties of feed seawater, NF permeate and NF retentate streams.[22]

Parameter	Feed seawater	NF permeate	NF retentate
pH	7.86	7.10	7.55
Conductivity ($\mu S \cdot cm^{-1}$)	56,600	45,600	73,200
Alkalinity as $CaCO_3$ ($mg \cdot L^{-1}$)	121	36	111
TDS ($mg \cdot L^{-1}$)	46,700	31,860	73,250
Total hardness ($mg \cdot L^{-1}$)	7,700	330	20,800
Calcium ($mg \cdot L^{-1}$)	489	56	1,242
Magnesium ($mg \cdot L^{-1}$)	1,576	46	4,305
Sulphate ($mg \cdot L^{-1}$)	3,450	<2	11,040
Chloride ($mg \cdot L^{-1}$)	25,140	19,970	31,559
Bicarbonate ($mg \cdot L^{-1}$)	148	44	135

Feed flow $= 8.03\,m^3\,h^{-1}$; NF permeate recovery $= 63.4\%$; feed pressure $= 24\,kg\,cm^{-2}$.

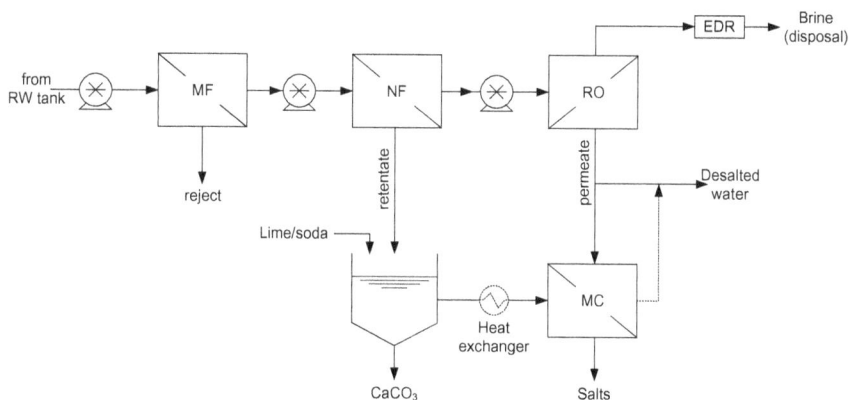

Figure 4.10. Flow chart of an integrated membrane system for the recovery of dissolved salts in NF retentate.

of vaterite at low supersaturation suggests that, because of the inhibiting effect of Mg^{2+} present in solution, the last two steps in the crystallization sequence, vaterite → aragonite → calcite, are retarded.

At high supersaturation, the rate of transformation of vaterite to calcite is very fast; under this circumstance, incorporation of magnesium ions in the crystalline lattice of calcite was observed.

Up to 89% of the initial content of calcium carbonate was removed; $35.5\,kg/m^3$ of NaCl and $8.4\,kg/m^3$ of $MgSO_4 \cdot 7H_2O$ per cubic meter of NF retentate were obtained. In addition, the amount of water condensed in the distillate side at the membrane crystallizer allowed the increase of the NF recovery factor from 64% to 95%.

For sodium chloride crystals, the values of the coefficient of variation, calculated from the experimentally obtained cumulative size distribution function, varied in the range of 15%–30% within the first hour of operation. Crystallization kinetics have been investigated under moderate supersaturation ratio ($S = 0.065 - 0.097$). As in a conventional industrial apparatus, primary heterogeneous nucleation and secondary nucleation have been found to play an important role in determining the crystallization kinetics. In particular, for a secondary nucleation rate given by

$$B = k_b M^j G^1. \tag{4.23}$$

Least square multiple linear regression analysis of kinetic data resulted in the following values: $k_b = 1.67 \times 10^{18}\,\text{s}^3 \cdot \text{m}^{-2} \cdot \text{kg}^{-1}$, $j = 0.8 \pm 0.3$, $i = 1.7 \pm 0.5$. The kinetic parameters k_g and g in Eq. (4.13) have been obtained from growth rate experiments as abscissa and slope from an $\ln(G)$ versus $\ln(\Delta c)$ plot, respectively. The linear regression, with a correlation coefficient of 0.90, gave $g = 1.17$ and $k_g = 1.92 \times 10^{-5}$ (G expressed in $\text{m} \cdot \text{s}^{-1}$). For Epsomite, kinetic results indicate a crystallization rate of about $5 \times 10^{-7}\,\text{m} \cdot \text{s}^{-1}$.

The mitigation of the environmental impact of discharged brines from RO desalination plants has inspired several technological options. Macedonio et al. (2011) proposed combing wind aided intensified evaporation (WAIV) and MCr in order to reduce the volume of concentrate and recovery salts from brackish water. WAIV is a method for making use of wind energy to evaporate a solution flowing on wetted surfaces packed in high density per footprint.

The integrated system was able to reach recovery factors as high as 88.9% and to dispose less than 0.27% of the raw feed water. NaCl crystals produced in MCr units showed the characteristic cubic shape and a linear growth rate of 0.0063 μm/min from RO retentate in the absence of antiscalant, of 0.0035 μm/min from RO retentate with antiscalant, a trace of organic compounds, and a 75% RO recovery factor.[29]

The investigation carried out by Chen et al. (2014) showed that feed flow rate and feed inlet temperature were the two dominant factors affecting the CSD. Specifically, an increase in the feed flow rate resulted in a smaller mean crystal size owing to a reduction in the crystallizer residence time. An increase in the feed inlet temperature had the same effect owing to an increase in the thermal load of the crystallizer and a corresponding decrease of the supersaturation (the driving force for crystal growth).[30]

(a)

(b)

Figure 4.11. Calcium carbonate crystals: (a) vaterite, (b) calcite polymorphs. (Reprinted from [9] with permission. Copyright © 2004 Elsevier B.V.)

4.5.2 *Reactive gas–liquid (G–L) membrane crystallization*

Carbon dioxide is the most critical greenhouse gas and its capture for reuse and storage is the objective of an extensive effort by many researchers. The crystallization of Na_2CO_3 by reactive absorption of CO_2 into alkaline NaOH solution, where the diffusion of gas is mediated by a G–L membrane contactor, has been recently investigated.[23,24]

In an absorbent alkaline liquid, CO_2 reacts with the hydroxyl ions according to the following scheme:

$$CO_2 + OH^- = HCO_3^- \qquad (4.24a)$$

$$HCO_3^- + OH^- = CO_3^{2-}. \qquad (4.24b)$$

The second-order reaction, Eq. (4.24a), limits the overall reaction rate.

In a G–L membrane crystallizer, the reactive absorption of CO_2 into alkaline solutions is carried out in a membrane contactor. By definition, the term *membrane contactor* is used to define a membrane system employed to keep in contact two phases using a microporous hydrophobic membrane as an inert interface. The membrane is not wetted by the polar phase, it can be filled by the gas phase, and mass transfer occurs prevalently by diffusion. A schematic repesentation of the process is illustrated in Fig. 4.12.

In order to provide a general discussion, the absorbing gas CO_2 is below indicated as a generic component A. The reactive absorption preliminarily involves the transfer of component A from the gas phase in the bulk to the membrane interface, then through the membrane, and finally into the liquid phase where the chemical reaction of A takes place.

The local flux of A, below indicated as J_A, is expressed in terms of the concentration gradient Δc_A (driving force) and it can be related to the serial resistances against mass transfer in the three phases:

$$J_A = \frac{1}{\frac{1}{k_g} + \frac{1}{k_m} + \frac{1}{mk_L E}} \Delta c_A, \qquad (4.25)$$

where m is the distribution coefficient and k_g, k_L, and k_m are the mass transfer coefficients in the gas, liquid, and membrane phase, respectively (Fig. 4.13).

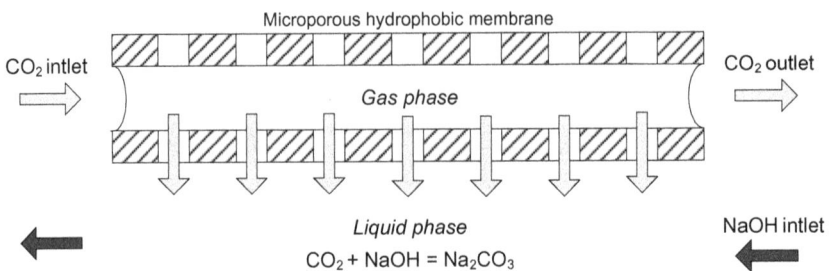

Microporous hydrophobic membrane

CO_2 intlet

Gas phase

CO_2 outlet

Liquid phase

$CO_2 + NaOH = Na_2CO_3$

NaOH intlet

Figure 4.12. Reactive absorption of carbon dioxide into sodium hydroxide solution occurring in a hollow fiber G–L membrane contactor.

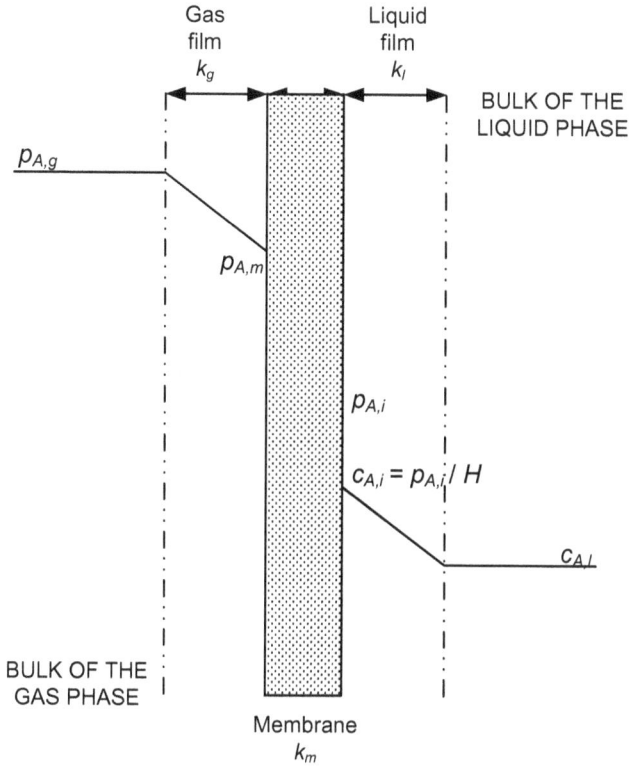

Figure 4.13. Film model for the absorption/reaction of the component A across a microporous hydrophobic membrane (p = partial pressure; H = Henry constant; c = concentration).[5]

It is convenient to introduce the enhancement factor E, defined as the ratio of the absorption flux of A in the liquid in the presence of a chemical reaction to the absorption flux of A without reaction:

$$E = \frac{J_A \mid \text{with reaction}}{J_A \mid \text{without reaction}}. \tag{4.26}$$

Depending on the specific system, the literature provides several approximate solutions to predict the value of E, based on different mass-transfer models. Common assumptions include the presence of a well-mixed and large liquid bulk zone adjacent to the diffusion l. For small diameters of the fibers, depending upon the gas-liquid contact time, the mass transfer zone in the liquid phase may extend up to the axis of the fiber.

In this respect, the Graetz number (Gz) is defined as the ratio of the penetration time of A to reach the axis of the hollow fiber to the average residence time of the liquid in the fiber:

$$Gz = \frac{v_L d^2}{D_A L}.$$
(4.27)

In Eq. (4.27), v_L is the liquid velocity, d is the diameter of the hollow fiber, D_A is the diffusion coefficient of the gaseous solute, and L is the fiber length.

At high values of Graetz number ($Gz \gg 1$), the penetration extent of A is small if compared to the internal radius of the fiber, and the average concentration of the gas in the lumen is very low. At $Gz > 1000$, the average bulk concentration is negligible in comparison with the gaseous solute concentration present at the G–L interface and the absorption process resembles, with good approximation, the one occurring into a liquid of infinite depth.[25]

The appropriateness of an approximate solution is generally limited to relatively high values of Gz, whereas rigorous numerical solutions are mandatory whenever operating at low Gz and for reactions having complex kinetics.

For the case of interest concerning the reactive absorption of a gas in an alkaline liquid flowing through a microporous hollow fiber membrane, the differential mass balance of the component A transferred is

$$v \frac{\partial c_A}{\partial z} = D_A \left[\frac{1}{r} \frac{\partial}{\partial r} \left(r \frac{\partial c_A}{\partial r} \right) \right] + \Psi,$$
(4.28)

where v is the velocity of the liquid along the fibers, c_A the concentration of component A, r the radial coordinate, and Ψ the chemical reaction rate. Equation (4.28) is derived under the following conditions:

(a) steady state and isothermal conditions;
(b) axial diffusion negligible;
(c) fully developed parabolic profile in the tube side;
(d) applicability of Henry law;
(e) cylindrical symmetry of hollow fibers.

Under laminar flow (this is an usual situation in hollow-fiber membrane modules), the velocity profile in a cylindrical conduct is parabolic (Fig. 4.14) and

$$v(r) = 2v_z \left[1 - \left(\frac{r}{R} \right)^2 \right],$$
(4.29)

where v_z is the average axial velocity and R the radius of the fiber.

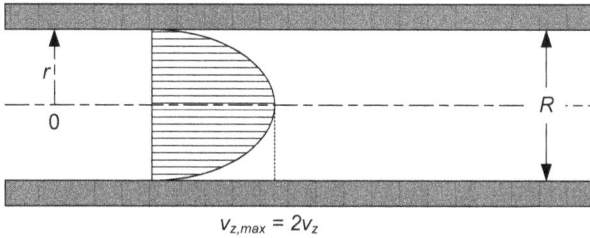

Figure 4.14. The parabolic profile of the axial velocity of a liquid flowing in a hollow fiber under laminar regime.

Equation (4.28) requires the following initial and boundary conditions in the axial and radial directions[26]:

- at $z = 0$ and $\forall\, r$, $c_A = c_{A,0}$;
- at $r = 0$ and for $z > 0$, $\left(\frac{\partial c_A}{\partial r}\right) = 0$ (symmetry);
- at $r = R$ and for $z > 0$, $\left(\frac{\partial c_A}{\partial r}\right)_{i \neq A} = 0$

  (it is assumed that all components are non-volatile, with the exception of the gaseous solute A);

- $D_A \left(\frac{\partial c_A}{\partial r}\right) = k_{est}(c_{A,g,\text{bulk}} - c_{A,g,\text{interface}})$

  (it is assumed that, at the membrane-liquid interface, the flux of component A in the liquid phase is equal to the flux in the gas phase).

In the last equation, D_A is the diffusion coefficient of the solute A. Subscript g refers to the gaseous phase. The external mass transfer coefficient k_{ex} is defined as

$$k_{ex} = \frac{1}{\frac{1}{k_g} + \frac{1}{k_m}}, \tag{4.30}$$

where k_g and k_m are the mass transfer coefficient in the gas and membrane phases, respectively.

Case 1. First-order irreversible reaction

For a first-order irreversible reaction, the reaction term in Eq. (4.28) is

$$\Psi = k_1 c_A, \tag{4.31}$$

where k_1 is the reaction kinetic constant.

Under the approximation of a mass transfer coupled to a first-order irreversible chemical reaction with infinite bulk, an asymptotic solution for the enhancement

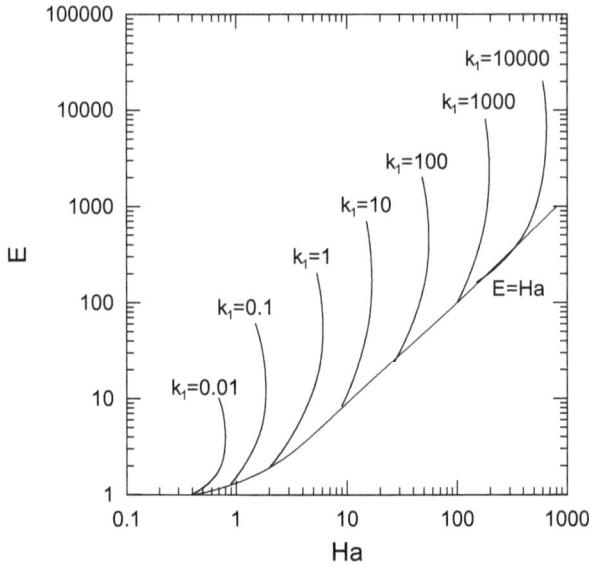

Figure 4.15. Plot of E vs. Ha for the gas absorption with first-order irreversible reaction in a liquid flowing through a hollow fiber membrane module. Simulation conditions: $L = 0.1\,\text{m}$; $d = 600 \times 10^{-6}\,\text{m}$; $c_{Ag} = 10\,\text{mol}\cdot\text{m}^{-3}$; $v_L = 1 \times 10^{-4} - 5.0\,\text{m}\cdot\text{s}^{-1}$; $k_1 = 1 \times 10^{-2} - 1 \times 10^4\,\text{s}^{-1}$; $D_A = 1 \times 10^{-9}\,\text{m}^2\cdot\text{s}^{-1}$; $m = 1$. (Reprinted from [5] with permission. Copyright © 2005 Elsevier B.V.)

factor E in a fast reaction regime based on surface renewal theory is given by

$$E = Ha = \frac{\sqrt{k_1 D_A}}{k_L}, \tag{4.32}$$

where Ha is the Hatta number that compares the maximum rate of reaction to the maximum rate of transport of A through the liquid film.

Figure 4.15 plots the enhancement factor E versus the Hatta number (Ha); here, the approximate asymptotic solution based on the surface renewal theory (straight line) is compared to the exact numerical results of Eq. (4.28) for the case of gas absorption in a liquid flowing through a hollow fiber accompanied by a first-order reaction.

The Ha number is varied by changing the first-order reaction rate constant k_1 and mass transfer coefficient k_L in the range of values reported in the figure caption. At high Graetz number, the enhancement factor calculated numerically approaches the predicted value of E, and the mass transfer process occurs in a thin zone adjacent to the G–L interface.

For a slow reaction regime ($Ha < 0.3$), there is no enhancement due to chemical reaction, and the absorption flux depends on the mass transfer coefficient and, hence, on the liquid velocity.

Case 2. Second-order irreversible reaction

The reaction rate for a second-order irreversible reaction is

$$\Psi = k_{1,1} c_A c_B. \tag{4.33}$$

Under the limiting conditions of high Graetz number, the penetration depth of the reacting species is allocated near the G–L interface. Moreover, it is assumed that the concentration of liquid phase reactant B at the axis of fiber is the same as the concentration of B at the inlet of the fiber (pseudo-first-order reaction regime). Under these conditions, the enhancement factor E is equal to the modified Ha[26]:

$$E = Ha = \frac{\sqrt{k_{1,1} c_{B0} D_A}}{k_L}, \tag{4.34}$$

where c_{B0} is the mixing cup concentration of B.

At sufficiently low Graetz number ($Gz \to 0$), i.e.

$$\frac{c_{B0} D_B}{c_{A,i} D_A} \ll Ha, \tag{4.35}$$

an instantaneous reaction regime can be assumed; the absorption rate is limited by the radial diffusion of the reacting species to the reaction plane and the flux is strongly influenced by the mass-transfer coefficient.

In this case, the asymptotic solution for the infinite enhancement factor E_∞ is given by

$$E_\infty = \left(1 + \frac{c_{B0} D_B}{v_B c_{A,i} D_A}\right) \left(\frac{D_A}{D_B}\right)^n, \tag{4.36}$$

where v_B is the stoichiometric coefficient of B. The value of n depends on the typology of the mass transfer model used; in a typical situation $n = 1/3$.

The diagram of E versus Ha for the case of a gas absorption in a liquid flowing through a hollow fiber accompanied by a second-order reaction (with $v_B = 1$) is illustrated in Fig. 4.16. The comparison between Figs. 4.15 and 4.16 shows that a maximum value for the enhancement factor (E_∞) is reached in the case of a second-order reaction for all $k_{1,1}$, whereas this limit is not present for a first-order reaction.

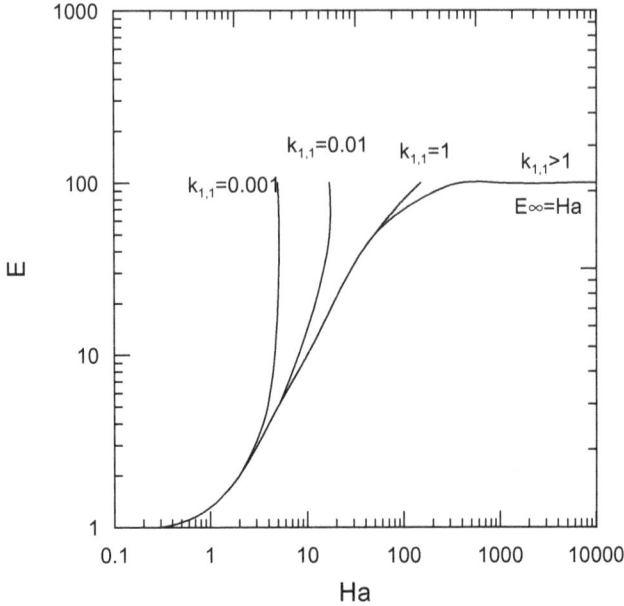

Figure 4.16. Plot of E vs. Ha for the gas absorption with second-order irreversible reaction in a alkaline liquid flowing through a hollow fiber model contactor. Simulation conditions: $L = 0.1\,\text{m}$; $d = 600 \times 10^{-6}\,\text{m}$; $C_{Ag} = 10\,\text{mol}\cdot\text{m}^{-3}$; $C_{B0} = 1000\,\text{mol}\cdot\text{m}^{-3}$; $v_L = 1 \times 10^{-4} - 5.0\,\text{m}\cdot\text{s}^{-1}$; $k_{1,1} = 1 \times 10^{-2} - 1 \times 10^{4}\,s^{-1}$; $D_B = D_A = 1 \times 10^{-9}\,\text{m}^2\cdot\text{s}^{-1}$; $m = 1$. (Reprinted from [5] with permission. Copyright © 2005 Elsevier B.V.)

Ye $et\,al.$ (2013) carried out crystallization tests in the presence of $NaNO_3$, NaCl, and Na_2SO_4 acting as impurities.[24] It was observed that NO_3^- and Cl^- impurities had no significant influence on the sodium carbonate crystal growth rate, whereas the presence of SO_4^{2-} significantly affected the shape of the crystals grown by evaporation in a conventional crystallizer, and triclinic crystals were obtained. On the other hand, the crystallographic analysis of Na_2CO_3 obtained by membrane-assisted crystallization (mostly in the deca-hydrate form) showed less sensitivity to the incorporation of impurities, and all crystals preserved the original prismatic structure. Markedly, the 0.06% impurity content of Cl^- in the crystals is much lower than that of Na_2CO_3 produced by the Solvay method (around 0.15%) and well below the upper limit for the production of glass (0.15%).

4.5.3 *Treatment of wastewater streams*

In 2004, Van der Bruggen and colleagues proposed the integration of NF, membrane distillation, and MCr for the advanced treatment of textile wastewater.[27] A typical

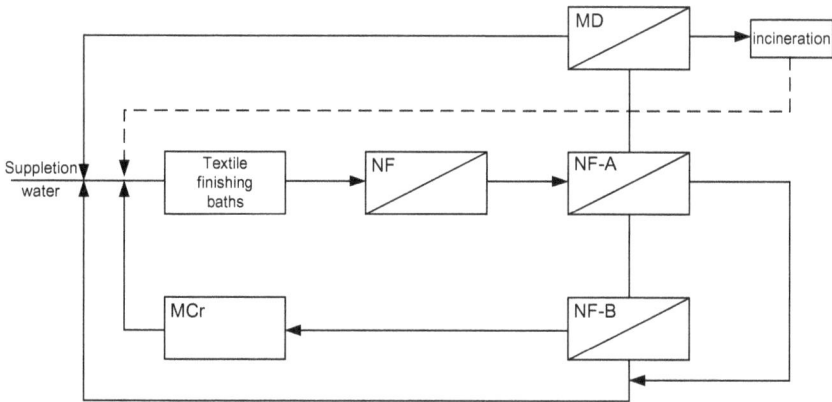

Figure 4.17. Integrated membrane system for wastewater treatment in the textile industry. (Adapted from [27] with permission. Copyright © 2004 Elsevier Ltd.)

process requires organic compounds in finishing baths, and salts to regulate the attachment of dyes or other compounds to the textile; moreover, finishing baths are often applied at elevated temperature (80°C–90°C), thus allowing waste heat exploitation.

According to the flow chart illustrated in Fig. 4.17, wastewater is pretreated by microfiltration (MF) for the removal of suspended solids; the organic fraction present in the MF permeate is then removed by a loose NF membrane with low salt rejection at a high temperature (NF-A). The permeate fraction of NF-A, containing the organics not retained and a large fraction of salts, is fed to a second NF unit (NF-B) whose retentate is sent to a membrane crystallizer for recovery of high purity crystals to be reused directly for new finishing baths. The warm retentate fraction of NF-A, rich in organic compounds, is fed to a membrane distillation unit for further concentration; the distillate is then recycled as fresh water.

The paper was revisited in 2013 to monitor the progress in this emerging field of research;[28] it was pointed out that membrane contactor technology still suffers limitations due to the scarce availability of specialty membranes on the market, cost issues, and lack of proof of concept on a large scale.

References

1. Hollander, E.D., Derksen, J.J., Bruinsma, O.S.L. *et al.* (2001). A numerical study on the coupling of hydrodynamics and orthokinetic agglomeration, *Chem. Eng. Sci.*, **56**, 2531–2541.

2. Kramer, H.J.M. and Jansen, P.J. (2002). Tools for design and control of industrial crystallizers, *Proc. of the 15th International Symposium on Industrial Crystallization*, Sorrento, Italy.

3. Genck, W.I. (2004). Guidelines for crystallizer selection and operation, *Chem. Eng. Prog.*, **100/10**, 26–32.

4. Curcio, E., Criscuoli, A. and E. Drioli. (2001). Membrane crystallizers, *Ind. Eng. Chem. Res.*, **40/12**, 2679–2684.

5. Drioli, E., Criscuoli, A. and E. Curcio. (2006). 'Membrane contactors: Fundamentals, applications and potentialities', in *Membrane Science and Technology Series 11*, Elsevier, Amsterdam, The Netherlands.

6. Ji, X., Curcio, E., Al Obaidani, A. *et al.* (2010). Membrane distillation crystallization of seawater reverse osmosis brines, *Sep. Pur. Tech.*, **71/1**, 76–82.

7. Curcio, E., Ji, X., Quazi, A.M. *et al.* (2010). Hybrid nanofiltration-membrane crystallization system for the treatment of sulfate wastes, *J. Membr. Sci.*, **360**, 493–498.

8. Klug, D.L. and Pigford, R.L. (1989). The probability distribution of growth rates of anhydrous sodium sulfate crystals, *Ind. Eng. Chem. Res.*, **28/11**, 1718–1725.

9. Drioli, E., Curcio, E., Criscuoli, A. *et al.* (2004). Integrated system for recovery of $CaCO_3$, NaCl and $MgSO_4 \cdot 7H_2O$ from nanofiltration retentate, *J. Membr. Sci.*, **239**, 27–38.

10. Jancic, S.J. and Grootscholten, P.A.M. (1984). *Industrial Crystallization*, Delft University Press, Delft, Holland, pp. 77–93.

11. Randolph, A.D. and Larson, M.A. (1988). *Theory of Particulate Processes*, Academic Press, New York, USA.

12. Tun, C.M., Fane, A.G., Matheickal, J.T. *et al.* (2005). Membrane distillation crystallization of concentrated salts — flux and crystal formation, *J. Membr. Sci.*, **257**, 144–155.

13. Wang, L., De Yoreo, J.J., Guan, X. *et al.* (2006). Constant composition studies verify the utility of the Cabrera–Vermilyea (C–V) model in explaining mechanisms of calcium oxalate monohydrate crystallization, *Cryst. Growth Des.*, **6/8**, 1769–1775.

14. Kubota, N. and Mullin, J.W. (1995). A kinetic model for crystal growth from aqueous solution in the presence of impurity, *J. Crystal Growth*, **152**, 203–208.

15. Westin, K.-J. and Rasmuson, A.C. (2005). Crystal growth of aragonite and calcite in presence of citric acid (CIT), DTPA, EDTA and pyromellitic acid, *J. Colloid Interface*, **282**, 359–369.

16. Curcio, E., Ji, X., Di Profio, G. *et al.* (2009). Membrane distillation operated at high seawater concentration factors: Role of the membrane on CaCO3 scaling in presence of humic acid, *J. Membr. Sci.*, **346/2**, 263–269.

17. Edwie, F. and Chung, T.-S. (2013). Development of simultaneous membrane distillation–crystallization (SMDC) technology for treatment of saturated brine, *Chem. Eng. Sci.*, **98**, 160–172.

18. Macedonio, F., Quist-Jensen, C.A., Al-Harbi, O. *et al.* (2013). Thermodynamic modeling of brine and its use in membrane crystallizer, *Desalination*, **323**, 83–92.

19. Pitzer. K. (1973). Thermodynamics of electrolytes I. Theoretical basis and general equations, *J. Phys. Chem.*, **77**, 268–276.

20. MEDINA Project (FP6 n. 036997) (2009). Progress Report D.1.3.2: Incorporate Pitzer activity model in solubility calculations of different scalants mainly $CaCO_3$ and $CaSO_4$.

21. Macedonio, F. and Drioli, E. (2010). Hydrophobic membranes for salts recovery from desalination plants, *Desalin. Water Treat.*, **18**, 224–234.
22. Drioli, E., Curcio, E., Criscuoli, A. *et al.* Integrated system for recovery of $CaCO_3$, NaCl and $MgSO_4 \cdot 7H_2O$ from nanofiltration retentate, *J. Membr. Sci.*, **239**, 27–38.
23. Luis, P., Van Aubel, D. and Van der Bruggen, B. (2013). Technical viability and exergy analysis of membrane crystallization: closing the loop of CO_2 sequestration, *Int. J. Greenh. Gas Control*, **12**, 450–459s
24. Ye, W., Lin, J., Shen, J. *et al.* (2013). Membrane crystallization of sodium carbonate for carbon dioxide recovery: Effect of impurities on the crystal morphology, *Cryst. Growth Des.*, **13**, 2362–2372.
25. Dindore, V.Y., Brilman, D.W.F. and Versteeg, G.F. (2005). Hollow fiber membrane contactor as a gas–liquid model contactor, *Chem. Eng. Sci.*, **60**, 467–479.
26. Kreulen, H., Smolders, C.A., Versteeg, G.F. *et al.* (1993). Microporous hollow fibre membrane modules as gas–liquid contactors. Part 2. Mass transfer with chemical reaction, *J. Membr. Sci.*, **78**, 217–238.
27. Van der Bruggen, B., Curcio, E. and Drioli, E. (2004). Process intensification in the textile industry: The role of membrane technology, *J. Environ. Manage.*, **73/3**, 267–274.
28. Van der Bruggen, B. (2013). Integrated membrane separation processes for recycling of valuable wastewater streams: Nanofiltration, membrane distillation, and membrane crystallizers revisited, *Ind. Eng. Chem. Res.*, **52**, 10335–10341.
29. Macedonio, F., Katzir, L., Geisma, N. *et al.* (2011). Wind-aided intensified evaporation (WAIV) and membrane crystallizer (MCr) integrated brackish water desalination process: advantages and drawbacks, *Desalination*, **273**, 127–135.
30. Chen, G., Lu, Y., Krantz, W.B. *et al.* (2014). Optimization of operating conditions for a continuous membrane distillation crystallization process with zero salty water discharge, *J. Membr. Sci.*, **450**, 1–11.

Chapter 5

Morphology, Polymorphism, and Co-Crystallization of Molecular Compounds

5.1 Introduction

In classical terms, crystallization from solution can be considered a three-step mechanisms: nucleation, growth, and growth cessation,[1] as schematized in Fig. 5.1. These steps are governed by supersaturation, and the way in which each of them develops, is associated with several macroscopic crystal properties.[2] When spontaneous crystallization starts, the process can follow several pathways that inevitably result in the formation of crystals of various sizes, size distribution, shapes, purity, and structure (polymorphism); these pathways depend on the relative rates of the nucleation and the growth stages. In general, the rates of nucleation and growth increase with supersaturation at a given temperature. At high supersaturation, nucleation is the dominant mechanism, showing proportionality with the extension of the crystalline population. Furthermore, in the nucleation stage, molecular orientation and aggregation mechanisms are intimately associated with the early selection of relative organization of growth units in the crystalline lattice. Therefore, crystal density and polymorphism are features related directly to the nucleation stage.

Once nuclei are formed, they can evolve into macroscopic crystals by one of the following methods: keeping their structural properties, dissolving to give rise to more thermodynamically stable aggregates, or experiencing phase change due to internal reorganizations or to liquid-mediated transitions. At this stage,

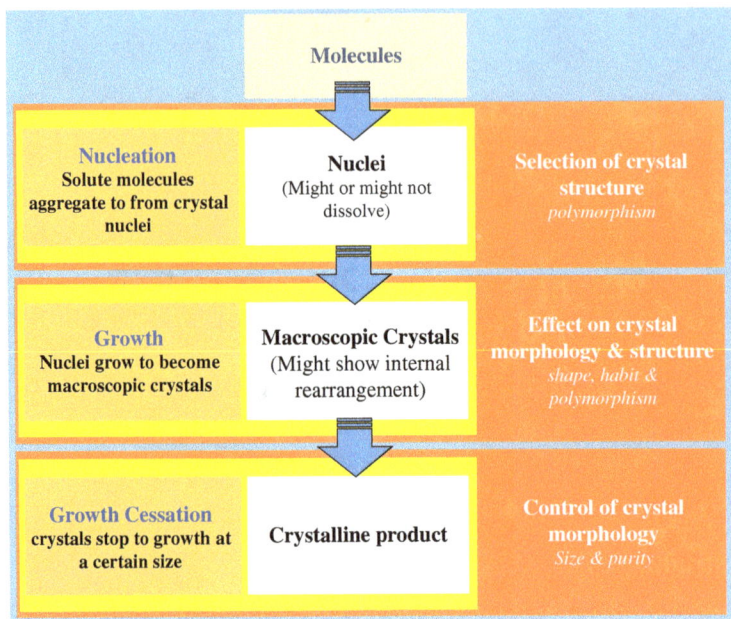

Figure 5.1. The steps of a crystallization process. Each stage might affect one or more final product properties.

the competition between the growth and the transformation rates has a direct influence on the final polymorphic composition of the precipitates. Additionally, crystal morphology can be influenced by the conditions of growth, e.g. by the solvent, the presence of impurities, fluid dynamics regime, etc., so that the external look (shape and habit) of the crystals depends on the way in which the growth units are supplied to and integrate into specific crystal faces.

Finally, crystals grow up to a certain size, which depends on the availability of growth units, the incorporation of impurities in the crystalline lattice, and the presence of additives which may adsorb to growing crystal faces thereby impeding the growth units to be included in the lattice. The occurrence of the growth cessation at a certain time, affects the size and the solvent inclusion (purity) in the crystalline lattice, since overly fast growth may lead to excessive liquid inclusions.

The description above highlights the role of supersaturation degree and generation rate on the competition between thermodynamic and kinetic factors in crystallization processes. As a result, product characteristics are strong function of the supersaturation change.[3] This clarifies the importance of supersaturation control in the crystallization processes in achieving the desired and reproducible features in terms of crystal number density, size, size distribution, shape, polymorphism, and purity.

5.2 The Metastable Zone

Theoretically, supersaturation at any level can cause nucleation of a crystal-free solution. However, primary (homogeneous) nucleation from solution does not usually occur spontaneously until the concentration exceeds a certain threshold value, as represented by the supersolubility curve in Fig. 5.2.[4] The region between the solubility and supersolubility curves is named the metastable zone, where spontaneous nucleation is not likely to occur but where heterogeneous nucleation stimulated by exogenous surfaces (impurities or dust particles), the crystal interfaces (secondary nucleation), or impeller/vessel surfaces can easily occur.

The metastable limit, in contrast to the saturation limit, is thermodynamically not defined.[5] It depends on a number of parameters such as temperature, supersaturation rate, impurities, fluid dynamics, etc.[6] Therefore, the metastable zone width (MZW) results from the specific properties of the crystallization process and can be considered to be system-specific.

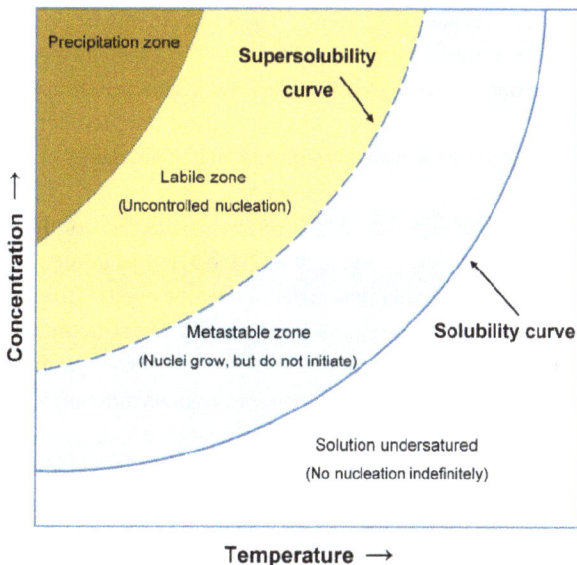

Figure 5.2. Phase diagram for a crystallizing system showing the stable, metastable, and unstable zones.

In the labile zone, nucleation starts stochastically and, as a result, product quality varies distinctively. For this reason, supersaturation needs to be maintained inside the metastable zone; an optimum supersaturation level is approximately half of the width of the metastable zone. Therefore, the knowledge of the metastable zone width is useful in defining the working point in the phase diagram of the crystallizing system, in order to design crystallization processes where the adaptability of working conditions during the process will allow the obtaining of specific crystal characteristics.

5.3 Crystal Morphology

With crystal morphology is meant the external appearance of particles, including shape, habit, size, and size distribution. Crystal morphology has a direct impact on downstream processes such as filtration, washing, drying, flow, compaction, dissolution, packaging, and storage.[7] Uncontrolled and unpredictable variations in crystal morphology can lead to a loss of control over products' features, often with costly results.[8] Although grinding or milling can be effective for providing more acceptable and reproducible properties, such treatments are often time-consuming and expensive. They may also introduce complications by promoting changes in crystal structure[9] and crystal perfection.[10] Therefore, in most industrial crystallization processes, habit modification is required to control both the production and the post-production properties of crystalline materials. This can be achieved by acting on some parameters such as cooling or evaporation rate, temperature, degree of supersaturation, selection of the solvent, and pH adjustment, or by deliberately adding additives that act as a habit modifier.[11]

Crystal shape is determined by the different growth rates of the different crystallographic faces. The slowly growing faces have the most influence on the growth shape. The growth of a particular face is defined by the crystal structure, the density of active growth sites, and environmental conditions. The structure of solid surfaces on a scale of nanometers to micrometers is the governing factor in habit modification. Supersaturation plays a dominant role in the determination of crystal shape. Therefore, it is possible to modify the crystal shape by dynamically controlling the supersaturation level during crystal growth.[12]

Figure 5.3 shows the different crystalline habits of paracetamol (acetaminophen) crystals obtained at different supersaturation regimes in membrane-assisted crystallization.[13] These morphologies are observed for various rates of solvent evaporation and correspond to those obtained conventionally at low, moderate, and high supersaturation (Fig. 5.4).[14]

Figure 5.3. Morphologies of monoclinic paracetamol grown at: (a) low, (b) medium, and (c) high flow rate by MAC. (Adapted with permission from Di Profio, G., Tucci, S., Curcio, E. *et al.* (2007). Controlling polymorphism with membrane-based crystallizers: application to form I and II of paracetamol, *Chem. Mater.*, **19(10)**, 2386–2388. Copyright 2014 American Chemical Society).

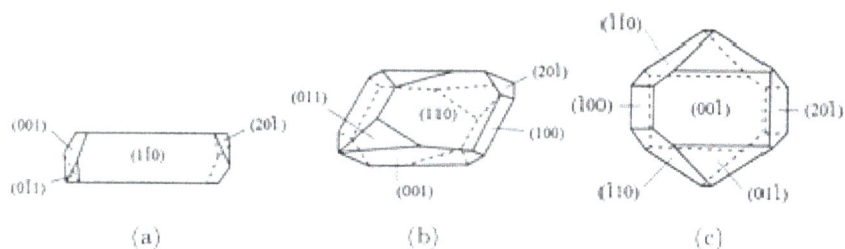

Figure 5.4. Morphology of monoclinic paracetamol grown at: (a) low, (b) medium, and (c) high supersaturation in conventional solution crystallization. (Adapted with permission from Ristic, R.I., Finnie, S., Sheen, D.B. *et al.* (2001). Macro- and micromorphology of monoclinic paracetamol grown from pure aqueous solution, *J. Phys. Chem. B*, **105**(38), 9057–9066. Copyright 2014 American Chemical Society).

5.4 Polymorphism

5.4.1 *Definitions*

A crystal polymorph, as defined by McCrone,[15] is a solid crystalline phase of a given compound resulting from the possibility of at least two different arrangements of the molecules in the solid state. While the term polymorph is reserved to crystals of equal chemical composition, the term pseudo-polymorphism is used to describe the ability of crystals of a certain substance to incorporate variable quantities of solvent with a defined stoichiometry.[16] Polymorphs arise because of the equilibrium resulting from the balance between attraction and repulsion among electron

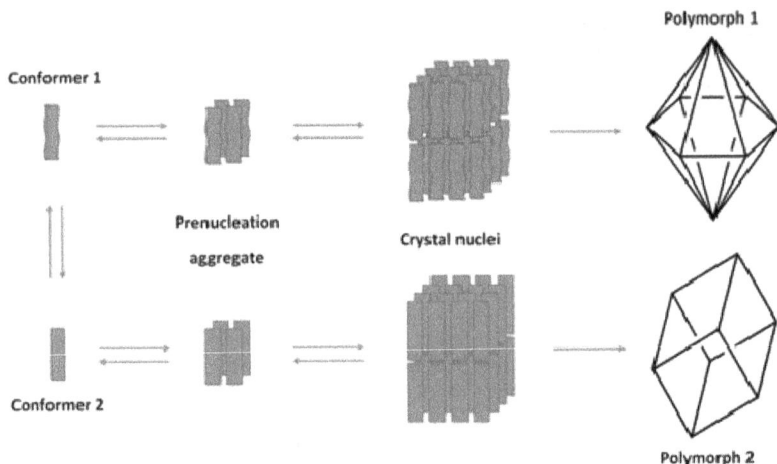

Figure 5.5.   Generation of different conformational polymorphs of the same molecule.

distributions of molecules in crystals. For molecular solids dominated by van der Waals interactions, the most energetically stable structure should be that which is most efficiently packed, i.e. most dense (the close packing principle). When other interactions, such as hydrogen bonds, dominate the packing, large voids in the crystal can be created, leading to energetically favorable, lower density structures.

Conformational flexible molecules introduce an intramolecular dimension to polymorphism, giving rise to structures that differ not only in the mode of packing, but also in molecular conformation, thus generating conformational polymorphs[17] (Fig. 5.5).[4] Computational studies suggest that many packing arrangements of organic molecules exist within approximately 8 kJ/mol to 10 kJ/mol of the most stable form, although most of these are not observed in practice.[18]

5.4.2   Phase diagram of polymorphs

Phase diagrams of different crystal phases are understood as diagrams of the solubility as a function of temperature and/or solvent composition. Figure 5.6 gives a schematic phase diagram for a dimorphic system with two modifications, named *I* and *II*. Two possibilities exist: (i) either one modification is stable at all temperatures below the melting point, so that *I* and *II* form a *monotropic* pair, and (2) the stability of *I* and *II* cross at a certain temperature below the melting point, as *I* and *II* form an *enantiotropic* pair.[16]

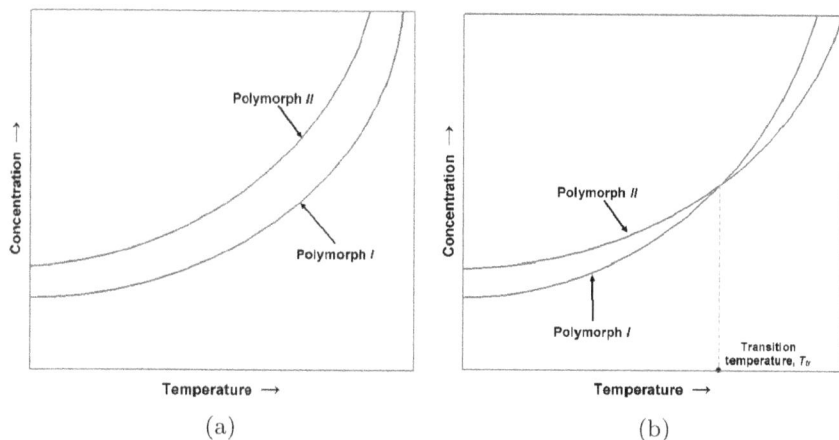

Figure 5.6. Schematic phase diagram of a system having two polymorphs, *I* and *II*. (a) A system where the modification *I* is the most stable at all temperatures below the melting point, (b) a system where the stability between *I* and *II* cross at a transition temperature below the melting point.

The solubility is directly related to the free energy of dissolution $\Delta_d G$ of the polymorph under consideration:

$$-RT \ln x_2 = \Delta_d G = \Delta_d H - T \Delta_d S, \tag{5.1}$$

where $\Delta_d H$ and $\Delta_d S$ depend on the solvent. The difference in free energy between polymorphs is

$$\Delta_{tr} G_{I/II} = \Delta_d G_I - \Delta_d G_{II}. \tag{5.2}$$

It is independent of the solvent and is equal also to the transition in the solid state. The enthalpy of phase transformation $\Delta_{tr} H_{I/II}$ is directly related to the difference in solubility:

$$\Delta_{tr} H_{I/II} = \Delta H_I - \Delta H_{II} = \frac{d \ln \chi_{II} - \ln \chi_I}{d^{1/T}}. \tag{5.3}$$

Consequently, differences in solubility directly relate to the differences in free enthalpy of the polymorphs. Since the ΔG between polymorphs is independent of solvent and directly proportional to the logarithm of their solubility ratio (assuming the activity coefficients approximately equal), by Eqs. (5.1)–(5.3) follows that the polymorph with the lowest solubility is the thermodynamically stable one. Therefore, the ΔG curves of monotropic polymorphs never intercept, while those of enantiotropic polymorphs intersect at a certain *transition temperature* (T_t), as shown in Fig. 5.7. In this latter case, since the enthalpy of *II* is lower than that of *I*, a quantity of energy $\Delta H_{tII/I}$ is required to be input for the phase transition, which

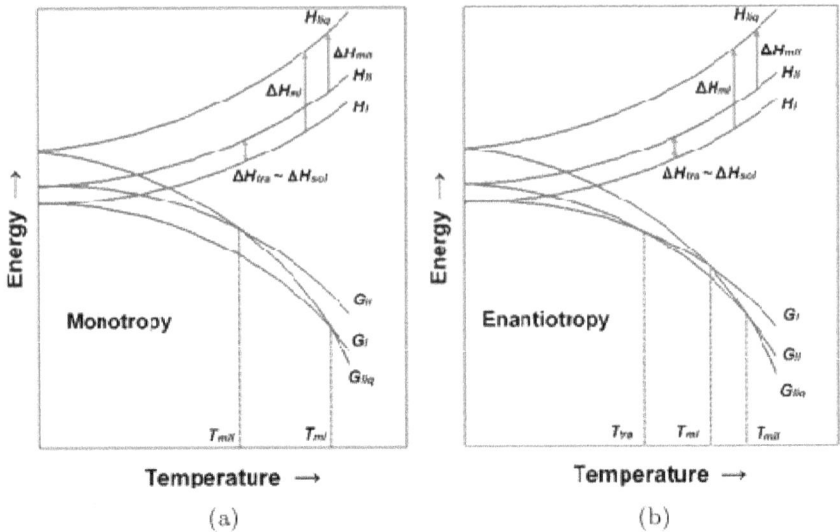

Figure 5.7. Energy-temperature diagrams for: (a) monotropically and (b) enantiotropically related crystal polymorphs of a dimorphic system.

must be endothermic for the transition above T_t and exothermic below T_t for the situation depicted in Fig. 5.7b.

The phenomenological manifestation of enantiotropism is that there can be a reversible transition from one phase to another without going through the gas or liquid phase.[19] As only when a solution can be supersaturated with respect to a single polymorph, can the production of the most stable crystal form *a priori* be assured. Therefore, knowledge of the enantiotropic or monotropic nature of the relationship between polymorphs can be useful in steering crystallization processes in order to obtain a desired polymorph at the exclusion of an undesired one.[20] This has been termed the *occurrence domain*.[21]

In regions in which there is an overlap of domains, one may expect that two or more polymorphs would crystallize, leading to concomitant polymorphism. Because the occurrence domains of enantiotropic polymorphs overlap over a range of temperatures above and below the thermodynamic T_{tra}, transition temperature does not represent a definitive value above and below which the high- and low-melting forms, respectively, can be crystallized. Even in monotropic systems, the metastable zones of the polymorphs may overlap, resulting in a common occurrence domain.

At thermodynamic transition points, two forms have the same stability and hence coexist as mixtures at equilibrium. At any other temperature there will be a thermodynamic tendency to transform to the more stable structure. This implies that mixtures of polymorphs will have limited lifetimes (except at T_{tra}) with transformation kinetics playing a role in those lifetimes.[22] Thus, although thermodynamics holds that only one crystal form will be stable at a given temperature and pressure, polymorphs may, depending on the kinetics of their crystal nucleation and growth, be concomitantly or exclusively formed.[16] The distinction between thermodynamic and kinetic influences is often demonstrated using the example of the graphite and diamond forms of carbon. The former is the thermodynamically preferred crystalline form, but kinetic factors (in particular a high activation barrier) make the rate of transformation from diamond to graphite infinitely slow.[23]

5.4.3 *Polymorph selection*

Ostwald's rule of stages

"When leaving a given state and in transforming to another state, the state which is sought out is not the thermodynamically stable one, but the state nearest in stability to the original state". This must be the next least stable state. This sentence represents an enunciation of *Ostwald's step rule* or *rule of stages*,[24] which states that the first form to crystallize is the one with the largest Gibbs free energy (the most soluble or the least stable), which then transforms to the next most soluble form (involving the smallest loss of free energy) through a process of dissolution and re-crystallization.

Ostwald's rule does not apply in all cases,[25] probably because the crystallization outcome is affected by many parameters, including solvent, cooling and stirring rates, temperature, pressure, and presence of impurities. Furthermore, crystal nucleation more frequently occurs heterogeneously on exogenous surfaces in contact with the solution.[26]

Kinetic control in polymorphs appearance

A practical implication of Ostwald's rule is that by manipulating the level of supersaturation, different polymorphs can be isolated. For a given polymorphic system to nucleate, a 3D nucleation barrier is to be overcome by any polymorph. This can be seen in the traditional energy/reaction coordinate diagram in Fig. 5.8.[16]

This shows G_0, the free energy per mole of a solute in a supersaturated fluid, which transforms by crystallization into one of two crystalline products, *I* or *II*. Associated with each reaction pathway are a transition state and an activation free energy. If a metastable form appears, its nucleation barrier is expected to be lower than that of the stable polymorph, which means that if *II* is the more stable,

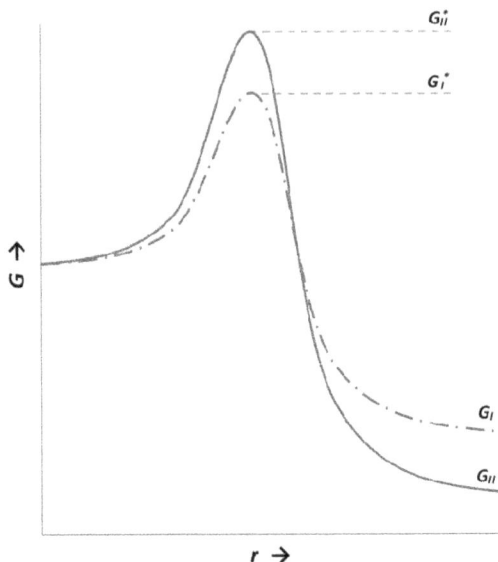

Figure 5.8. Schematic representation of the reaction coordinate r for crystallization in a dimorphic system to show the activated barriers for the formation of polymorphs I and II.

$G_I > G_{II}$. On the basis of classical nucleation theory (CNT), the magnitudes of the activated barriers are dependent on the size of the supramolecular assembly (crystal nucleus). The higher the operating level of supersaturation, the smaller this size is. In Fig. 5.8, the supersaturation with respect to I is $G_0 - G_I$ and is lower than $G_0 - G_{II}$ for structure II ($S_I < S_{II}$). However, if the critical size is lower for I than for II for a particular solution composition, then the activation free energy for nucleation is lower and kinetics will favor form I ($G_I^* < G_{II}^*$). Ultimately, form I will have to transform to form II according to thermodynamics. Therefore, the finite lifetime of the metastable polymorph is a result of the height of the kinetic barrier for this transformation. In the case of a small kinetic barrier, the transition can take place in the solid state. In the case of polymorphic systems involving large molecules, the barrier is often so high that this transition is rather difficult and far from instantaneous. If the system is still in contact with the solution, then the transition is expected to be faster, for example, via dissolution/recrystallization.[27] The solvent mediated phase transformation has two main rate limiting steps:[28] (i) the nucleation of the new phase and (ii) a combination of the processes of dissolution of the old phase, transport to the new phase, and its growth. The dissolution and crystallization rates depend on the solubility of both polymorphs, the actual concentrations, and the kinetics of the dissolution of the metastable polymorph

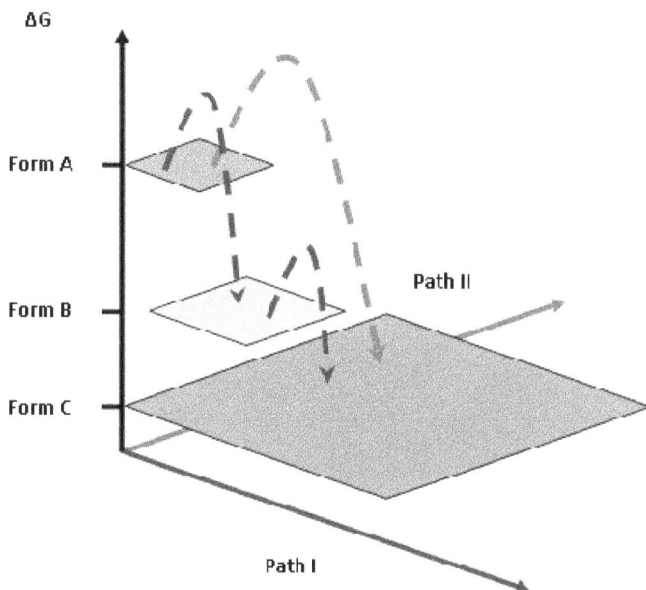

Figure 5.9. Schematic representation of the path of a solution mediated phase transformation from an unstable form A to the — thermodynamically — stable form C. Two pathways are possible, either directly to the stable form or via a second metastable form B. The path is determined by the nucleation.

and crystallization of the stable one. The polymorph generated via the solution mediated phase transformation is not necessarily directly the stable one; instead, a second, metastable form can occur (Fig. 5.9).

Related to the nucleation barrier is the MZW. The higher the nucleation barrier, the larger the metastable zone width of the polymorph is. Compared to the case of homogeneous nucleation, this width can be lowered if a polymorph nucleates heterogeneously on surfaces or interfaces that are in contact with the solution, even including the surfaces of another polymorph of the same substance already present in the solution.[29]

Considering a monotropic, dimorphic system (for simplicity) whose solubility diagram is shown schematically in Fig. 5.10,[16] it is quite clear that for the occurrence given by solution compositions and temperatures that lie between the form *I* and *II* solubility curves, only polymorph *II* can crystallize. However, the outcome of an isothermal crystallization that follows the crystallization pathway indicated by the vector in Fig. 5.10 is not so obvious since the initial solution is now supersaturated with respect to both polymorphic structures, with thermodynamics favoring form *II*, but kinetics form *I* in accordance with the Ostwald's rule.

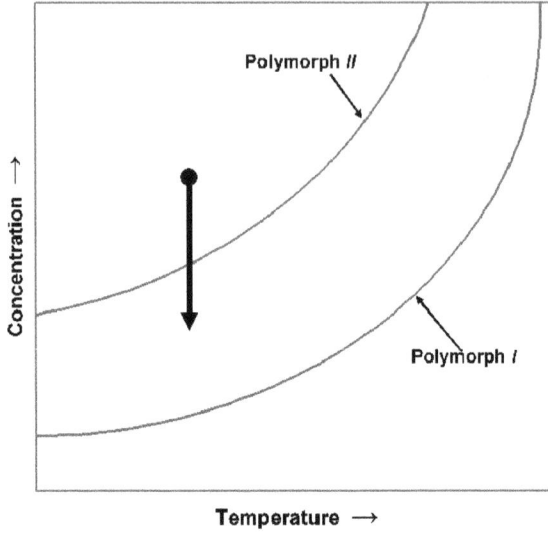

Figure 5.10.   Schematic solubility diagram for a dimorphic system (polymorphs *I* and *II*), showing a hypothetical crystallization pathway at constant temperature.

Influence of the degree of supersaturation on polymorphic selection

According to CNT, the influence of the degree of supersaturation on nucleation can be interpreted by considering the ratio of the nucleation rates of two polymorphs *I* and *II*, as given by Eq. (5.4), provided that the pre-exponential factor, A_n, is identical for the two polymorphs.

$$\frac{J_I}{J_{II}} = \exp \frac{16\pi \upsilon^2}{3(k_B T)^3} \left(\frac{-\gamma_I^3}{(\ln \sigma_I)^2} + \frac{\gamma_{II}^3}{(\ln \sigma_{II})^2} \right). \qquad (5.4)$$

Here, γ_I and γ_{II} represent the interfacial tensions of the *I* and *II* polymorphs, respectively, while σ_I and σ_{II} represent the supersaturation with respect to the corresponding polymorphs. The pre-exponential factor is essentially the product of the number of solute molecules in the solution and the probability that the solute molecules will aggregate to form a nucleus. If the solvent does not hinder the probability of aggregation of either polymorph, A_n is expected to be identical for the *I* and *II* polymorphs. The γ value of each of the polymorphs at a certain temperature may be calculated using

$$\gamma = 0.414 k_B T (c^s N_A)^{\frac{2}{3}} \ln \left(\frac{c^s}{c^*} \right), \qquad (5.5)$$

where c^s is the molar density of the polymorph. The higher nucleation rate of polymorph I than of polymorph II may result from the lower interfacial tension of polymorph I. However, as the supersaturation increases, the effect of σ on the nucleation rate becomes more pronounced than the effect of interfacial tension, γ (Eq. (5.5)). At high σ, the σ of polymorph I approaches that of II, so the relative nucleation rate, J_I/J_{II}, approaches unity. Hence, there may be conditions in which this is the occurrence domain for the polymorphs that crystallize concomitantly.[30]

Effect of supersaturation on the growth rate

Unless the crystal growth rate is limited by diffusion of solute molecules from the solution bulk to the growth surface, the rate is limited by the surface adsorption and integration of the molecules into the crystal lattice. Generally, for low viscosity solutions, the diffusion from the solution bulk to the growth surface is assumed not to be a rate-limiting step. In this case, the crystal growth mechanism may follow the continuous model, the birth and spread model, or the Burton–Cabrera–Frank (BCF) model, depending on the roughness of the growing surface.[31]

According to the BCF model, the effect of σ on crystal growth rate, R, is given by[32]

$$R = B(\sigma - 1)^2 \tanh\left(\frac{D}{\sigma - 1}\right), \tag{5.6}$$

where B and D are complex temperature-dependent constants. Equation (5.7) expresses the ratio of the crystal growth rates of the I and II polymorphs, R_I and R_{II} respectively, provided that the value of B is identical for these polymorphs:

$$\frac{R_I}{R_{II}} = \left(\frac{\sigma_I - 1}{\sigma_{II} - 1}\right)^2 \frac{\tanh\left(\frac{D}{\sigma_I - 1}\right)}{\tanh\left(\frac{D}{\sigma_{II} - 1}\right)}. \tag{5.7}$$

The value of D depends on the step size for crystal growth and is typically on the order of 0.1.

Low values of supersaturation exert opposing effects on the rates of nucleation and crystal growth of polymorphs I and II. At low supersaturation, the relative nucleation rate, J_I/J_{II}, is considerably greater than the relative crystal growth rates, R_I/R_{II}, and hence the nature of the polymorph that crystallizes is determined by J_I/J_{II}. At high supersaturation, $J_I/J_{II} \to 1$, but R_I is still significantly higher than R_{II}. As a result, the II polymorph crystallizes at high supersaturation. Therefore, the competition between the rates of nucleation and crystal growth of the two polymorphs determines the nature of the polymorph that crystallizes.[33]

5.5 Polymorphism and Drugs

The increasing structural complexity of modern pharmaceutical agents may account for the increasing frequency of polymorphic behavior of their crystals. Polymorphism is a relevant phenomenon because the diverse structures of polymorphs make them unique, patentable materials.[34] Therefore, in recent years, the pharmaceutical industry has increasingly focused its attention on polymorphs because of the dramatic differences in the physico-chemical properties, like melting point, solubility, dissolution rate, optical properties, and compressibility,[35] exhibited by different polymorphs, which can have a great deal of influence on the bioavailability, filtration, tableting, *etc.*[36]

The impact of crystal polymorphism on pharmaceutical product value has been illustrated by costly product failures and protracted patent litigation examples. Emblematic of the importance of polymorphs was the notable example of Norvir® (Ritonavir), the AIDS drug made by Abbott Laboratories. Introduced in 1996, the drug had been on the market for 18 months when suddenly, during manufacturing, a previously unknown polymorph, thermodynamically more stable than the drug in its original form, was detected. Although the two polymorphs shared a chemical formula, the second form was only half as soluble as the first, so patients taking prescribed doses would not get enough of the drug into their bloodstreams.[37] The Company could not find a way to stop the formation of the new polymorph. Within a few days of its discovery, this new polymorph was dominating the product coming off the lines. Abbott pulled Ritonavir from the market and decided to reformulate the drug in the second polymorphic form as a liquid gel capsule containing the pre-dissolved drug. Unlike the original formulation, the gel capsules require refrigeration. The company spent hundreds of millions of dollars trying to recover the first polymorph and lost an estimated $250 million in sales the year the drug was withdrawn.

On the other hand, a newly discovered polymorph may turn out to be a more effective and convenient than the original product. Early entry of generic versions of some notable drugs has been enabled based on the use of patently distinct but pharmaceutically equivalent polymorphs.[38] Ranitidine HCl (Zantac®),[39] the anti-ulcer drug owned by GlaxoSmithKline, is a prominent example where patent strategy and gaming created significant uncertainty about the value of a drug franchise. A process resulting in the crystallization of form I was patented in 1978. Two years later the more stable crystalline form II appeared which was also patented. In 1995, when the form I patent expired, other companies began gearing up to market cheaper, generic versions. However, GlaxoSmithKline still had a patent on the second polymorph of the drug. Because the second form was easier to manufacture, it became the active ingredient in Zantac. Although other companies were legally permitted to make and sell generic versions of the first polymorph of ranitidine hydrochloride, they had to figure out how to make it without any contamination from the second, whose patent protection remained in force. This kept the generic products off the market for several years.

In theory, all pharmaceuticals are vulnerable to these unexpected events, which in the case of Ritonavir and Ranitidine HCl resulted in enormous cost and inconvenience for the innovator and impacted the patients' use. Therefore, in response to these challenges, regulatory agencies require the evaluation of the impact of polymorphism on product performance[40] and require that pharmaceutical manufacturers demonstrate that each polymorph is stable and can be reproduced reliably. As different polymorphs can translate into more or less profit, processes and techniques for the identification and characterization of polymorphs and solvates of compounds are of interest and have been developed.[41]

5.6 Current Strategies for Obtaining the Desired Polymorph

In 1965 McCrone stated the following maxim: "The number of forms known for a given compound is proportional to the time and money spent in research on that compound". This is currently known as McCrone's law.[15] More than half of compounds can crystallize in more than one polymorphic form, either as true polymorphs or solvates and hydrates.[42] In many cases, a process yields crystals of more than one polymorph at the same time, making the control of polymorphism even more challenging.[16]

Controlling which polymorph will emerge in the crystallization of pharmaceuticals requires selecting between competing crystallization pathways that are governed by thermodynamics, crystal nucleation and growth kinetics, and molecular recognition events.[43] The crystallization process of polymorphous crystals is composed of competitive nucleation, growth, and the transformation from a metastable to a stable form.[44] The schematic diagram of the controlling factors in crystallizations of polymorphs is shown in Fig. 5.11.[33] The basically influential factors are denoted as primary factors, while additives,[45,46] solvents,[47] and interfaces[48] are included in the secondary factors. Typical techniques used to control the growth of a desirable polymorph are based on pressure conditions, addition of impurities, surfactants, modification of the viscosity of the solution, use of electric/magnetic fields, etc. However, the mechanism of these effects is not known and the quantitative relationship between the operational factors and the crystallization characteristics of polymorphs is not clearly understood and, in many cases,

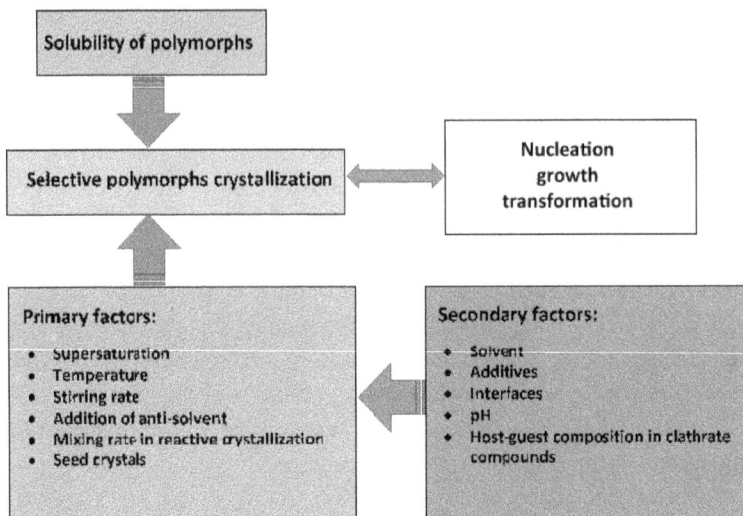

Figure 5.11. Scheme of controlling factors in crystallization of polymorphs.

is system specific. Temperature and supersaturation are considered to be basically important as controlling factors.[49]

5.7 Controlling Polymorphism by Membrane-Assisted Crystallization

5.7.1 *Glycine*

Glycine is a model system because it has three polymorphs: α, β, and γ.[50–52] Two additional forms, δ and ε, were obtained under high-pressure conditions.[53] Among the first three forms, β-glycine is the least stable and crystallizes from water/alcohol mixtures.[54] Form β is monotropically related to the other two phases α and γ, which are enantiotropically correlated each other. The two modifications α and γ are known to grow from pure aqueous solutions at different ranges of pH: α-glycine appears from almost neutral solutions ($4 < \mathrm{pH} < 8$) while the γ-modification can be formed from acidic or basic solutions ($\mathrm{pH} < 4$; $\mathrm{pH} > 8$). This behavior has been initially associated to the different speciation of molecules in the different pH ranges; while glycine exists in solution as cyclic dimers of the zwitterionic form[53] around the isoelectric point ($\mathrm{pI} = 4.96$), for extreme pH values the molecules are in the form of charged species which inhibit the growth of the α form, giving an increased chance that the γ form will grow. Although this

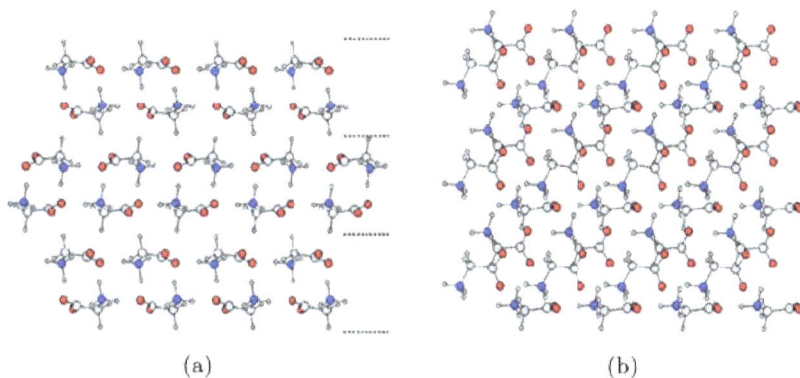

(a)   (b)

Figure 5.12.   (a) The crystal structure of α-glycine viewed along the a-axis, shown in green and the interface between hydrogen bonded sheets shown as black, dashed lines. (b) The crystal structure of γ-glycine viewed along the a-axis, with hydrogen bonds shown in green.[53]

description has been questioned, the different speciation in solution is considered as the origin of the different structure of the two phases (Fig. 5.12) so that the α form shows a bi-dimensional structure in which bi-layers are linked together by hydrogen bonds, while the γ-glycine consists of molecules forming a three-dimensional network.[53]

The relative stability at ambient temperature of the three forms was found to be $\gamma > \alpha > \beta$.[55] This is an indication that the relative stability of the polymorphs does not correlate directly with the ease of their crystallization, and that the spontaneous nucleation of α-glycine in aqueous solutions is kinetically rather than thermodynamically controlled.

Different techniques have been used to obtain the selective production of the γ-form from neutral aqueous solutions. These are the use of additives to inhibit the growth of α-glycine,[56] the exposure of supersaturated solutions to intense pulses of plane-polarized laser light,[57] the confinement on patterned substrates[58] and the crystallization in emulsions and lamellar phases.[59]

The different three-dimensional structures of the two forms cause dissimilarities in their typical vibrational spectra and thermal behavior. Differences in the vibrational band frequencies of each solid phase can consequently be attributed to variations of the intermolecular hydrogen bonding in each crystal structure. From Fourier transformed infrared in attenuated total reflectance mode (ATR–FTIR) the two forms α and γ show distinct types of spectra. An example of each one of the two spectra is shown in Fig. 5.13.[36,60]

The differential scanning calorimetry (DSC) of the two crystal forms is also useful to identify them (Fig. 5.14).[61] At room conditions the order of stability

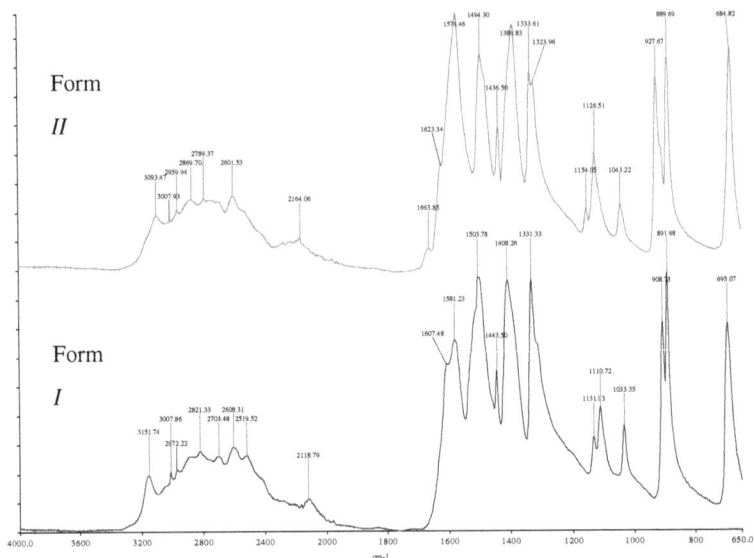

Figure 5.13.   ATR-FTIR spectra of the two crystal forms α and γ of glycine (Adapted with permission from Di Profio, G., Tucci, S., Curcio, E. *et al.* (2007). Selective glycine polymorph crystallization by using microporous membranes, *Cryst. Growth Des.*, **7(3)**, 526–530. Copyright 2014 American Chemical Society).

between the two polymorph of glycine is $\gamma > \alpha$. The $\gamma \rightarrow \alpha$ transition is thermo-dynamically forbidden at this temperature while the reverse transformation $\alpha \rightarrow \gamma$, which should be thermodynamically allowed, is not generally observed because of kinetic hindrance. As the two polymorphs of glycine are enantiotropically related, a certain temperature $T_{trans\gamma/\alpha}$, exists at which the α form becomes more stable than the γ modification, thanks to an entropic contribution to the free Gibbs energy G ($G_\alpha \leq G_\gamma$ and $H_\alpha > H_\gamma$, but $S_\alpha > S_\gamma$). Hence, at $T \geq T_{trans\gamma/\alpha}$ a phase change, in which the γ polymorph converts in the α modification, is observed. This is an endothermic transition, due to the difference $\Delta H_{\alpha/\gamma}$ between H_γ and H_α, as shown in the first heating run in Fig. 5.14b.

The polymorphic conversion temperature is known to depend strongly on the size and shape of the crystals, on the thermal pre-treatments, and on grinding and other mechanical treatment of the crystals.[62] Generally, for a ground glycine sample a large signal is observed, caused by the block-after-block transformation of the γ form into the α modification.[61] In Fig. 5.14b this transition is evidenced by a wide signal between 163°C and 175°C with two other weak signals around to 150°C and 185°C. By cooling down the α form below the $T_{trans\gamma/\alpha}$ (the first cooling run in Fig. 5.14b), no reverse transformation $\alpha \rightarrow \gamma$ is observed ($\gamma \rightarrow \alpha$

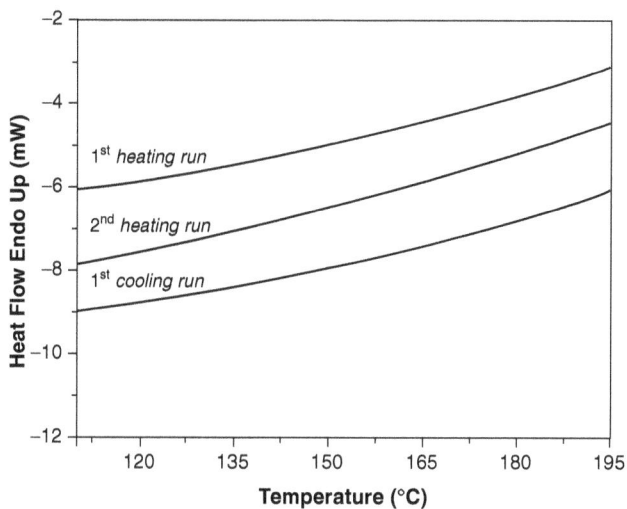

(a)

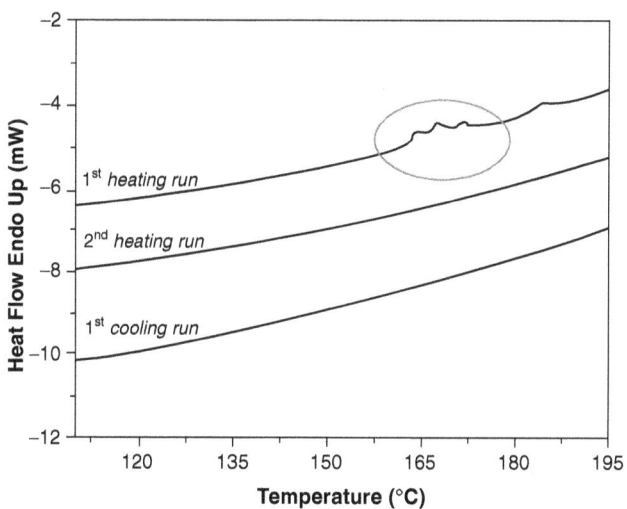

(b)

Figure 5.14. Representative DSC thermographs of the two forms of glycine: curve (a) Form α; curve (b) form γ. (Adapted with permission from Di Profio, G., Tucci, S., Curcio, E. *et al.* (2007). Selective glycine polymorph crystallization by using microporous membranes, *Cryst. Growth Des.*, **7(3)**, 526–530. Copyright 2014 American Chemical Society).

Table 5.1. Glycine polymorphs obtained in different conditions of solvent extraction flow rate (J) in the static configuration, and at various recirculation solutions velocities (v) in the dynamic system in membrane-assisted crystallization (Adapted with permission from Di Profio, G., Tucci, S., Curcio, E. *et al.* Selective glycine polymorph crystallization by using microporous membranes, *Cryst. Growth Des.*, **7(3)**, 526–530. Copyright 2014 American Chemical Society).

J (mL/h)	Polymorph	v (μm·sec^{-1})	Polymorph
6.9×10^{-3}	γ	349.6	γ
13.8×10^{-3}	γ	539.9	γ
18.4×10^{-3}	α	709.1	α
20.7×10^{-3}	α	899.9	α
23.0×10^{-3}	α	989.9	α
27.6×10^{-3}	α	1349.8	α
34.5×10^{-3}	α	1845.5	α

is irreversible), because of the high activation energy. In the subsequent heating run (the second heating run in Fig. 5.14b), no transitions are observed as the complete conversion of the γ-glycine to the α form occurred in the first heating stage. No such transformation is observed for the α-structure, as shown in Fig. 5.14a.

Table 5.1 reports the polymorphic outcomes of membrane-assisted crystallization of glycine at pH 6.2. It can be seen that the preferential crystallization of γ-glycine depends on the solvent evaporation rate through the pores of the membrane in a static MCr system at pH 6.2, where the α form is expected to be grown. For $J < 1.38 \times 10^{-2}$ mL/h the only form obtained is γ-glycine, while for $J \geq 1.84 \times 10^{-2}$ mL/h the kinetic product, α-glycine, always appears.

The curve in Fig. 5.15 clearly shows that the value of supersaturation, in correspondence to which the nucleation occurs, increases with the rate of solvent evaporation. Namely, the supersaturation at the nucleation point increases from 1.19 for $J = 6.9 \times 10^{-3}$ mL/h to 2.41 for an evaporation rate of 3.45×10^{-2} mL/h.

In a dynamic membrane-assisted crystallization, a similar polymorphic selectivity is observed depending on the solution circulation velocity (v). For recirculation solution velocities lower than $500 \, \mu$m·sec^{-1} the γ-polymorph was selectively crystallized, while for v higher than $700 \, \mu$m·sec^{-1}, the metastable α-form is the only one obtained (Table 5.1). In a dynamic membrane crystallizers, the effect of the solution velocity impacts directly on the solute concentration profile nearby the membrane surface and therefore on the transmembrane flow rate; J increases with v, as the mass transfer coefficients rise. Therefore, as the flow rate for solvent extraction (and, consequently, the generation rate of supersaturation) is high,

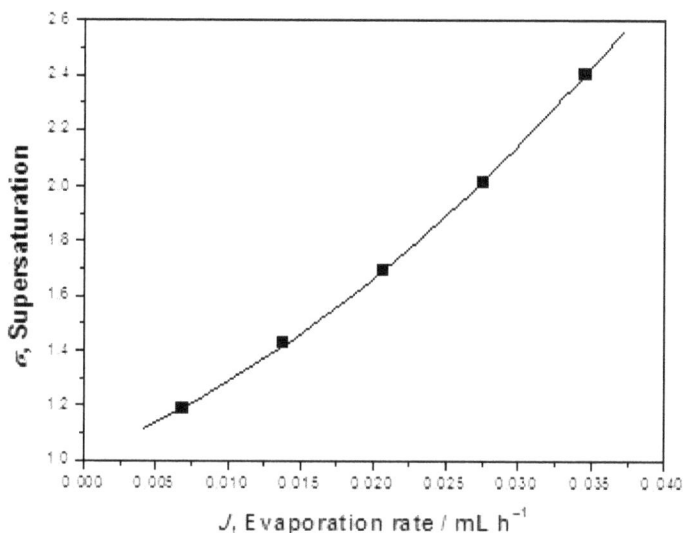

Figure 5.15. Relation between the solvent evaporation rate J and the supersaturation σ at the nucleation point for the static membrane-assisted crystallization. (Adapted with permission from Di Profio, G., Tucci, S., Curcio, E. *et al.* Selective glycine polymorph crystallization by using microporous membranes, *Cryst. Growth Des.*, **7(3)**, 526–530. Copyright 2014 American Chemical Society).

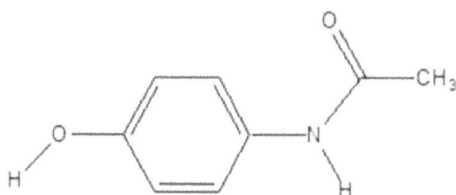

Figure 5.16. Paracetamol ($C_8H_9NO_2$).

the kinetically favored form appears, while for low rate of solvent extraction the growth of the more stable form is favored.

5.7.2 Paracetamol

Paracetamol (acetaminophen, Fig. 5.16) is a widely used antipyretic and analgesic drug.

Three polymorphs of paracetamol are known: the monoclinic[63] form *I* (space group $P2_1/n$) is the thermodynamically stable modification at room temperature;

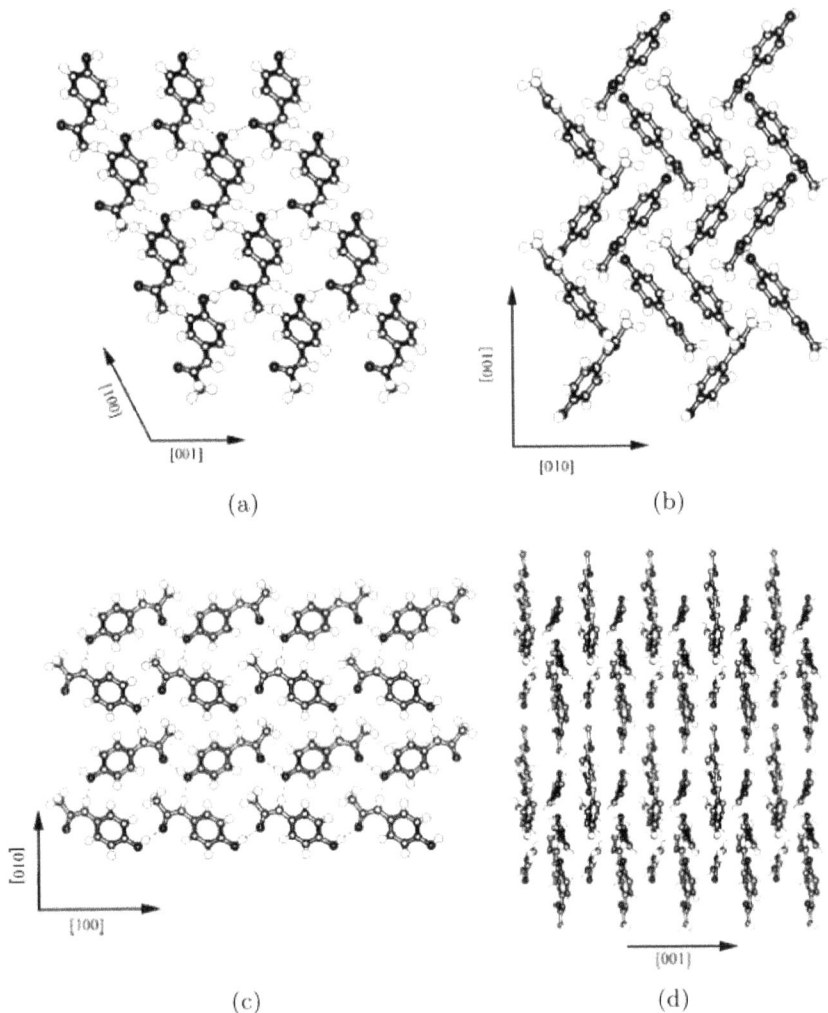

Figure 5.17. Projections of monoclinic paracetamol on plane: (a) (001) and (b) (010) and of the orthorhombic modification on plane: (c) (110) and (d) (001).[69]

form *II* (orthorhombic, space group *Pcab*)[64] is metastable at ambient conditions; form *III* is the least stable form that is obtained by crystallization from the melt.[65] The crystal structure of the marketed monoclinic form lacks a sliding plane; its crystal structure has pleated sheets stacked along the b-axis (Fig. 5.17), making it relatively stiff. This results in poor compression properties and therefore in the requirement of binding agents for tableting.

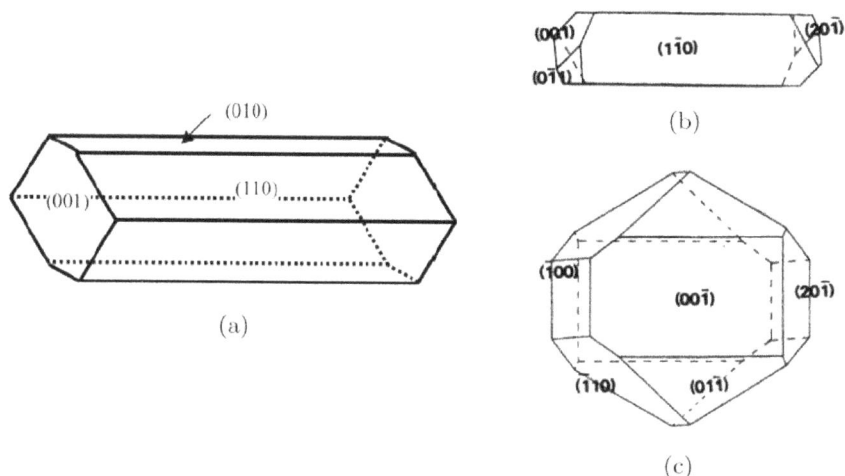

Figure 5.18. Paracetamol (a) form *II* and form *I* crystals grown at (b) low (5%–7%) and (c) high (15%) supersaturation.[70]

Although improved compression properties can be obtained by varying the habit of the crystallites,[66] such a practice has problems in completely eliminating the residues of the required solvent,[67] thus being time-consuming and with an increased production cost.[68] In this respect, much attention has been focused on the production of form *II*. In fact, the crystal structure of the orthorhombic form *II* (Fig. 5.18) has parallel hydrogen-bonded sheets along the c-axis, giving slip planes that allow plastic deformation.[69] Therefore, the direct compression of the orthorhombic form *II* into tablets has been investigated for potential industrial use.[70] This makes form *II* potentially more attractive from an industrial standpoint for the direct compression in tablets.

While form *I* can be easily grown from solutions in various solvents,[70] form *II* has been obtained only by complicated or multi-step processes.[67,70,71] Form *I* crystals grown from aqueous solution exhibit an elongated prismatic morphology at low supersaturation,[72] but for relatively high supersaturation the faces are equally developed, and so supersaturation plays a role in determining the habit (Fig. 5.18a).[73] This is a remarkable point as the morphology of paracetamol crystals also affects the industrial processing. For the monoclinic form, the experimental prismatic-to-plate-like morphology is well known (Fig. 5.18a).[70]

Paracetamol molecules in the crystal are linked in chains by NH···O and OH···O hydrogen bonds. The chains, in turn, are linked in a two-dimensional plane and form a layered three-dimensional structure with van der Waals interactions

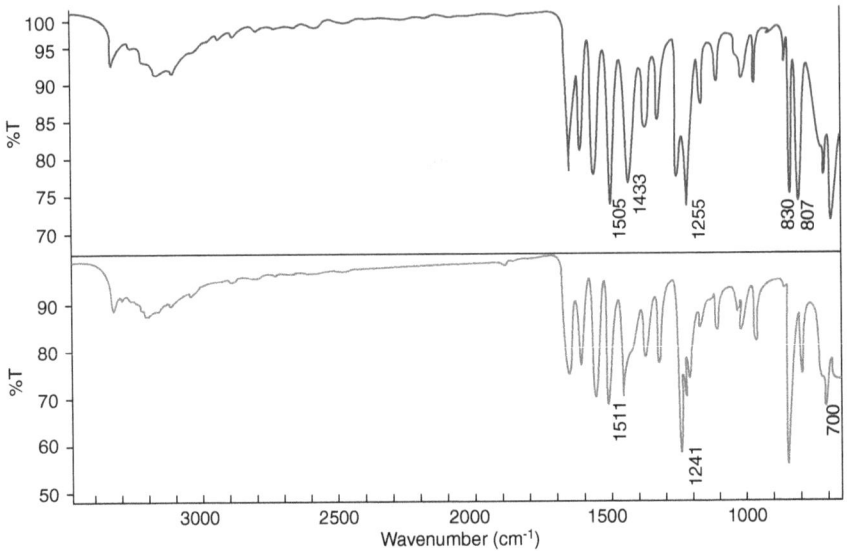

Figure 5.19.   ATR-FTIR spectra of the two polymorphs of paracetamol.

between the layers. The chains in the monoclinic and rhombic modifications are topologically similar but are bound in layers in different ways, as shown in Fig. 5.17.[74]

The shift of some specific peaks and some infrared absorptions patterns in the range 1700–650 cm^{-1} are the most useful in distinguishing between the monoclinic and orthorhombic form of paracetamol (Fig. 5.19)[75]; in the spectra of form *I* a peak around 1515 cm^{-1} and a strong absorption at 807 cm^{-1}, whilst in those of form *II*, absorptions around 1666 cm^{-1}, 1622 cm^{-1} and 1423 cm^{-1}.

The thermograph for form *I* shows a single strong endothermic event (Fig. 5.20a).[70] This sharp peak is due to the melting of form *I* at about 172°C (onset = 160°C). Form *II* gave a thermograph with three events, as shown in Fig. 5.20b. The first exothermic event, which is broad and weak, occurs over the temperature range 126°C to 140°C and is centered at about 133°C. The second event is a strong sharp peak at about 157°C (onset = 141°C). The third endothermic event is a weak sharp peak at about 171°C (onset = 168°C). The thermally induced events for form *II* (in order of increasing temperature) are a solid state conversion of form *II* to form *I*, followed by the melting of non-converted form *II*, and finally, the melting of form *I*.[76] This indicates that form *I* is polymorphically pure, whereas a solution-grown form *II* may have a low level of form *I* present.

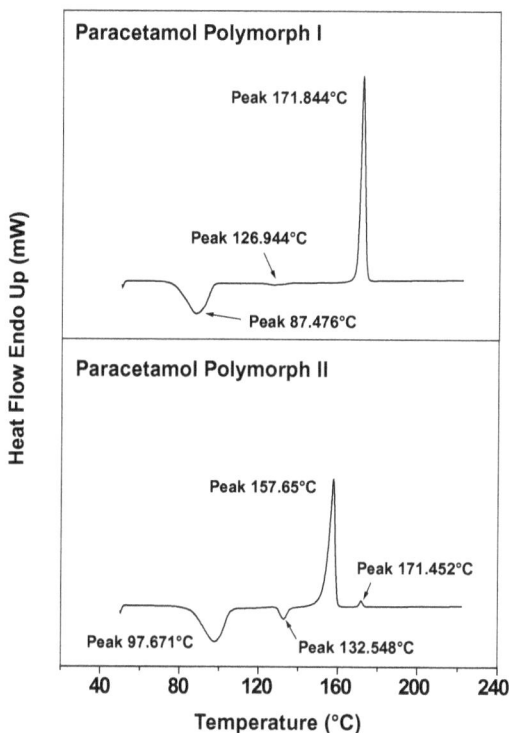

Figure 5.20.   DSC Thermographs of the two polymorphs of paracetamol: (a) form *I*; (b) form *II*.

In Table 5.2 the polymorphic yield in the membrane-assisted crystallization of paracetamol for different J is shown. The crystallization of one form was achieved by careful control of the solvent evaporation rate through the pores of the membrane. For $J \geq 7.9 \times 10^{-2}$ mL/h the polymorph obtained was the metastable form *II*, while for $4.33 \times 10^{-2} \leq J \leq 6.6 \times 10^{-2}$ mL/h the thermo-dynamic monoclinic product appeared. In the case of $J = 1.94 \times 10^{-2}$ mL/h, a weak peak in the spectra at about 808 cm^{-1} indicates the concomitant presence of a small amount of the modification *I* in the sample of form *II*.

Particular attention needs to be paid to Fig. 5.21, which induces consideration of different nucleation mechanisms for paracetamol under the conditions of very low J. In fact, the total rate of nucleation is the sum of several contributions (homogeneous, heterogeneous, surface, attrition-induced), one of which is dominant in certain ranges of supersaturation. The different slopes of the two straight lines fitting the experimental points identify a different nucleation mechanism in the diverse ranges of supersaturation: for high σ, a prevalently homogeneous

Table 5.2. Paracetamol polymorphs obtained at different conditions of rate of solvent evaporation (J), and corresponding induction time (t_{ind}) and supersaturation at the induction time ($\sigma°(t_{ind})$) (Reprinted with permission from Di Profio, G., Tucci, S., Curcio, E. (2007). Controlling polymorphism with membrane-based crystallizers?: application to form I and II of paracetamol, *Chem. Mater.*, **19(10)**, 2386–2388. Copyright 2014 American Chemical Society).

J (mL/h)	t_{ind} (h)	$\sigma°(t_{ind})$	Polymorph
1.37×10^{-2}	64.3	1.34	II
1.94×10^{-2}	53.6	1.52	II (+ I)
4.33×10^{-2}	23.5	1.91	I
4.37×10^{-2}	31.0	1.83	I
5.89×10^{-2}	24.3	1.93	I
6.60×10^{-2}	23.0	1.96	I
7.90×10^{-2}	19.5	2.00	II
8.98×10^{-2}	17.5	2.05	II

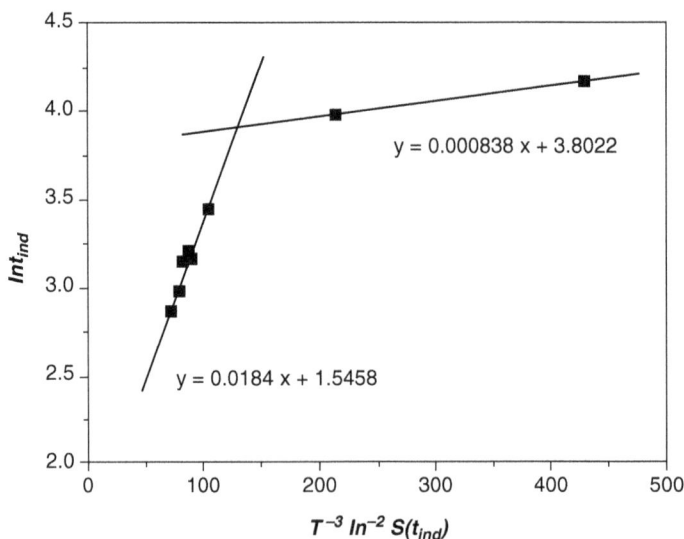

Figure 5.21. Behavior of ln (t_{ind}) as function of ($T^{-3} \cdot \ln^{-2} \sigma$). The diverse slopes of the two straight lines denote the different nucleation mechanisms ($T = 293\,K$). (Reprinted with permission from Di Profio, G., Tucci, S., Curcio, E. (2007). Polymorphism with membrane-based crystallizers: application to form I and II of paracetamol, *Chem. Mater.*, **19(10)**, 2386–2388. Copyright 2014 American Chemical Society).

nucleation, whereas for low σ a heterogeneous nucleation pathway.[77] Therefore, the presence of the polypropylene membrane surface provides very low values of $J (<2 \times 10^{-2}$ mL/h) heterogeneous nucleation sites, and hence an increased chance of nucleation for the orthorhombic form of paracetamol.[78,79] On the other hand, for higher J, the nucleation mechanism is predominantly homogeneous.

5.7.3 Carbamazepine

Development and scale-up of crystallization processes for active pharmaceutical ingredient (API) formulation in the form of metastable polymorphs is crucial since they can be utilized for a faster dissolution rate to achieve rapid absorption or a higher solubility to achieve acceptable exposure for a poorly water-soluble drug.[80] Carbamazepine (CBZ) is an example of a water-insoluble drug that has a high dose requirement (>100 mg/day) for therapeutic effect.[81] Since the solubility and dissolution rate of a thermodynamically more stable form can be expected to be lower than that of the metastable form, the possibility of producing higher energetic polymorphs is desirable, to attempt to improve the dissolution and adsorption.

It is known that CBZ exists in a dihydrate form (CBZ DH) and at least five anhydrous forms[82]: primitive monoclinic (CBZ III), triclinic (CBZ I), C-centered monoclinic (CBZ IV), and trigonal (CBZ II), sorted by thermodynamic stability in decreasing order,[82] and the orthorhombic form V.[83] In the crystallization of CBZ by membrane-assisted techniques, upon increasing the transmembrane flow rate, the amount of CBZ I in the precipitate decreases and the higher energetic form CBZ IV increases, as is shown in Fig. 5.22. Accordingly, the favored formation of metastable higher energetic forms is promoted at high rates of supersaturation. This result has also been rationalized by evaluating the thermodynamic and kinetic interplay in the crystallization process and it was concluded that the CBZ crystallization had been kinetically rather than thermodynamically controlled at high transmembrane flux.

The phase composition of the CBZ precipitates highlights the kinetic nature of the CBZ crystallization pathway. Indeed, solid samples were composed mainly of CBZ I and CBZ IV phases, while CBZ DH was only obtained as a byproduct. CBZ IV is a polymorph of low thermodynamic stability, which requires higher energy than CBZ I to be formed. On the other hand, CBZ I is the polymorph with lower symmetry and more populated asymmetric unit ($Z' = 4$), which can be obtained from the other phases upon heating up to $160°C-190°C$.[82] Crystal forms with more than one independent molecule in the asymmetric unit ($Z' > 1$) have often been referred to as kinetic products.[85] In fact, the $Z' > 1$ crystals represent a form which has been trapped before the molecules have adjusted themselves into

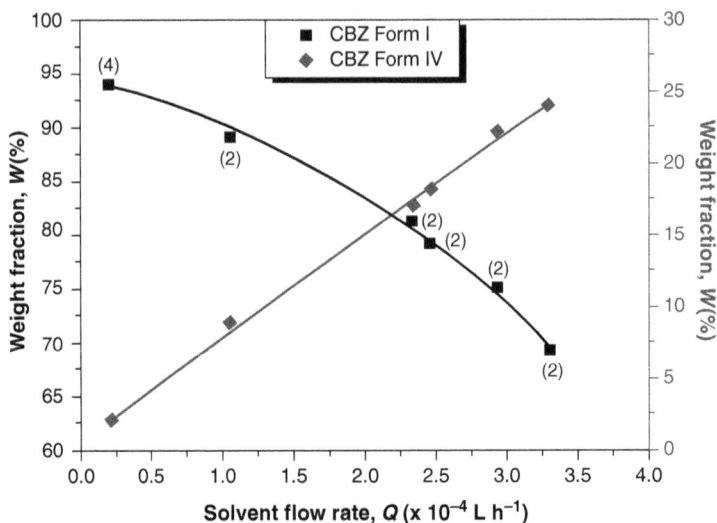

Figure 5.22. Weight fraction of CBZ polymorphs I and II vs solvent flow rate (in parentheses the weight fraction of the dihydrate form). Lines are only guides for the eyes. (Adapted with permission from Caridi, A., Di Profio, G., Caliandro, R. *et al.* (2012). Selecting the desired solid form by membrane crystallizers: crystals or cocrystals, *Cryst. Growth Des.*, **12**(7), 4349–4356. Copyright 2014 American Chemical Society).

a form with higher symmetry and $Z' = 1$. In other words, the $Z' > 1$ occurrence is related to the molecules' organization in stable clusters prior to reaching the highest symmetry arrangement in strong hydrogen-bonded structures.[86] It is a joint opinion that high Z' packing is due to kinetic trapping in the metastable state during the crystallization process. In this regard, high Z' structures have been referred to as "fossil relics",[87] a "snapshot picture of crystallization"[88] or "crystals on the way",[86] particularly when a lower Z' form also exists. Therefore, the preferential formation of CBZ polymorphs, either thermodynamically unfavored (CBZ IV) or with $Z' > 1$ (CBZ I) resulting from an interrupted crystallization, confirms that the CBZ crystallization mechanism occurrs under kinetic control in the explored transmembrane conditions.

5.7.4 *Effect of supersaturation control in membrane crystallization on polymorph selectivity*

When the thermodynamic and kinetic paths of crystallization are similar, the system has every chance to sample all the possible multi-molecular clusters in solution before selecting the one with the lowest energy. In this situation, the system is able to select the path that will minimize the energy at each stage (see Section 5.4.3).

Instead, when the crystallization process is under kinetic control, the system — leaving an unstable state — does not seek out the most stable state; rather, it finds the nearest metastable state which can be reached with loss of free energy, according to Ostwald's law of stages.[24] In this case, the system is localized into a relative energy minimum, leading to the formation of a metastable structure (kinetic product).

Depending on the rate of solvent extraction, the stable or a metastable polymorph can be selectively produced. This can be explained by considering the effect of supersaturation and the rate of its variation on the width of the metastable zone. Several studies have already demonstrated that the MZW increases with the rate of achievement of supersaturation.[91,92] The metastable limit normally increases with the rate at which the supersaturation is generated.[92] For a given polymorphic system, the crystallization of one form is associated with the overcoming of its nucleation barrier (this is related to the MZW). During nucleation, the magnitudes of the driving force for the aggregation of the solute molecule change with the surface area-to-volume ratio of the embryos and give rise to a critical radius above which the nucleus grows and below which it re-dissolves into the solution. This means that manipulation of the supersaturation rate might impact the occurrence domain of the different phases in a polymorphic system. When the rate of variation of supersaturation is low, if the least stable structure or a mixture of critical nuclei form at the same time, then nuclei of the more stable polymorph have time to grow at the expense of the less stable form, via solvent-mediated transfer of solute. However, for higher solvent evaporation rates, the increase in the MZW induces nucleation at higher values of supersaturation and the sudden achievement of such supersaturation leads to the nucleation and growth of the higher energetic form, which is the first to appear according to Ostwald's rule of stages.[24] However, although it is known that a less stable polymorph may nucleate first in solution due to a higher nucleation rate, it is energetically favorable for it to transform to a more stable polymorph over time.[93] Therefore, in many situations, it is difficult to isolate a metastable polymorph before it undergoes a solvent-mediated phase transformation to the more stable form.[94] This low-energy route exists because the molecules in solution possess translational, rotational, and torsional degrees of freedom, and can re-orient, reconfigure, and recrystallize to minimize the overall free energy of the system. For instance, this is the case for form *II* and, mainly, *III* of paracetamol, which tend to convert to the stable form *I*.

One possible route to obtain and stabilize a less stable polymorph from solution is to use specific additives to either encourage the growth of the desired form or disrupt the growth of the other form.[46] For instance, the very unstable form *III* of paracetamol was isolated by suppressing the molecular degrees of freedom by warming the amorphous glassy state and, then, imposing a large activation energy barrier to the interconversion between the different forms.[70] In many situations, heterogeneous nucleation is more energetically favorable than homogeneous nucleation and, therefore, tailor-made additives or substrates can be designed to manipulate the crystallization process toward a particular crystal structure.[95]

In MAC, transmembrane flux is directly connected with supersaturation rate. For low supersaturation rates, if the least stable structure or a mixture of critical nuclei forms at the same time, then nuclei of the more stable polymorph have time to grow at the expense of the less stable form, via solvent-mediated transfer of solute. For high solvent evaporation rates, the increase in the MZW induces nucleation at higher values of supersaturation and the sudden achievement of such supersaturation *freezes* the system in the production and growth of the metastable form, which is the first to appear according with the Ostwald rule of stages. It is hence straightforward the strictly relation between the MZW and the polymorphic yield of the system.

The membrane matrix acts as a selective barrier for solvent evaporation, allowing a progressive and finely controlled evaporation mechanism, which affects the MZW. In this last case, the increase of the supersaturation at the induction time by increasing J is evident, which indicates the increase in the MZW as J rises. Polymorph selection, independent of the solvent evaporation rate, is therefore directly related to the rate of achievement of the supersaturation necessary for nucleation and, in the end, to the MZW. This is responsible for the variation in the kinetics of the nucleation stage and, finally, of the switching from thermodynamic to kinetic control in the nucleation stage in the different conditions. The result is the selective production of either a stable or metastable polymorph as observed for glycine,[60] paracetamol,[13] and CBZ.[84]

This experimental evidence demonstrates that the driving force of the crystallization (supersaturation) and the rate of its variation can be effective tools in order to promote polymorphic selection during crystallization. As in a membrane crystallizer, the control of the solvent evaporation rate, by modulating the gradient of partial pressure (which is the driving force of the evaporation process), allows very precise control of the rate at which the supersaturation is reached, allowing switching from thermodynamic to kinetic control of the nucleation stage that turns into the selective nucleation of either the stable or the metastable polymorph.

5.8 Production of Pharmaceutical Co-Crystals

Co-crystallization of APIs with co-crystal formers (or co-formers) has gained significant interest in drug development, since the resulting new solid forms have different physico-chemical properties compared to the original API.[80,96,97] Co-crystallization increases the diversity of the solid-state forms of an API and enhances its pharmaceutical properties such as chemical stability,[96] moisture uptake,[98] mechanical behavior,[99] solubility,[100] dissolution rate, and bio-availability.[101] Therefore, co-crystallization can directly impact the scientific and legal aspects of drug development by providing alternative solid dosage forms and extended patent life.[80,97]

Co-crystallization is currently performed by a wide variety of techniques including solvent evaporation, dry and wet (solvent-drop addition) solid-state grinding, cooling, growth from melts, and slurry conversion;[102] among these, the first two are by far the most commonly used.[103] Co-crystal production requires special control during the supersaturation stage in order to avoid the over-running of the thermodynamic stability region in the phase diagram of a specific species, thus producing undesired or impure solid products.[41,96] Another major challenge concerns the scale up from laboratory to large scale without changing the optimized properties[80]; designing a scalable (co)crystallization process requires[104] the precise control of the working point within the phase stability diagram during the supersaturation stage, in such a way that the system is only supersaturated with respect to a particular phase.[103,104]

In this respect, MCr technology allows fine control of the evolution of the crystallization process, thus providing the opportunity to perform well-defined co-crystallization processes. The selective formation of carbamazepine-saccharin co-crystals (CBZ-SAC) from water/ethanol solvent mixtures, by using the appropriate initial solution composition, confirms this possibility.

5.8.1 *Dependence on initial solution composition*

Figure 5.23 displays the nature of the crystalline precipitate obtained in different test of MCr for several initial composition of the solution. It shows the occurrence domain of pure CBZ-SAC co-crystals, pure SAC and CBZ crystals, and mixtures of two crystalline products (CBZ and CBZ-SAC or CBZ-SAC and SAC). Precipitate characterizations revealed that products from experimental trials were extremely sensitive to solution composition. When decreasing the initial CBZ/SAC molar ratio from 1.1 to 0.01 the composition of the precipitate drastically changed, as depicted in Fig. 5.23.

CBZ crystals were obtained for CBZ/SAC molar ratio higher than 0.3; further decreasing the molar ratio in the range 0.3–0.22 produced a mixture of co-crystals/CBZ; CBZ-SAC cocrystals were obtained for solution composition consisting in 0.12–0.22 molar ratio, while a further decrease of molar ratio from 0.12

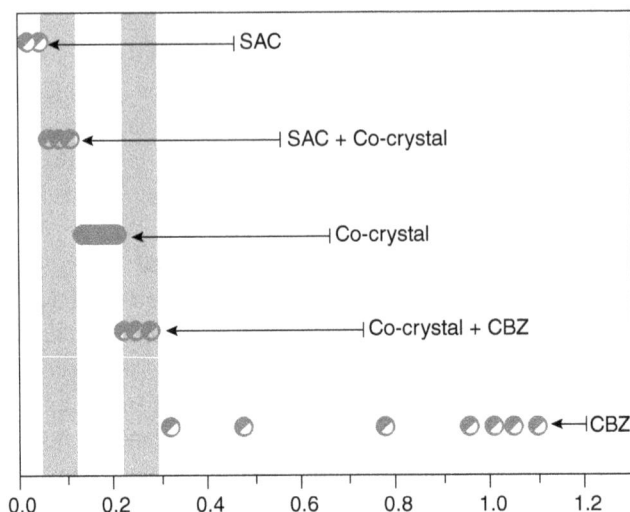

Figure 5.23. Zones of occurrence of the different products (indicated by the arrows) vs the initial CBZ/SAC molar ratio (Adapted with permission from Caridi, A., Di Profio, G., Caliandro, R. *et al.* (2012). Selecting the desired solid form by membrane crystallizers: crystals or cocrystals, *Cryst. Growth Des.*, **12(7)**, 4349–4356. Copyright 2014 American Chemical Society).

to 0.01 allowed the production first of co-crystals/SAC (up to 0.05) and then SAC crystals. These observations confirm the great importance of the initial solution composition in addressing the final outcome of the co-crystallization process and can be explained by exploring the phase solubility diagram of the studied system (Section 5.8.2).[105]

5.8.2 *Study of phase equilibrium diagram*

Since the crystal nucleation and growth depends on solution composition, temperature, and solvent evaporation rate in a rather complex way, examination of solution composition in the context of a phase diagram is the most effective means of showing how the preferential supersaturation for a specific component can be generated. The ternary phase diagram (TPD), which shows the solubility of co-crystal and pure components, is useful to give a detailed visualization of the system.[106, 107] Figure 5.24a displays a theoretical TPD for a congruently saturating system, where A (API) and B (co-former) components have similar solubility in the solvent S.[107] Here, the L region comprises undersaturated solutions and it is bounded by a and b to solubility curves of the two single components A and B, respectively. Thus, the intersection of these three curves provides the eutectic points (c_1 and c_2),

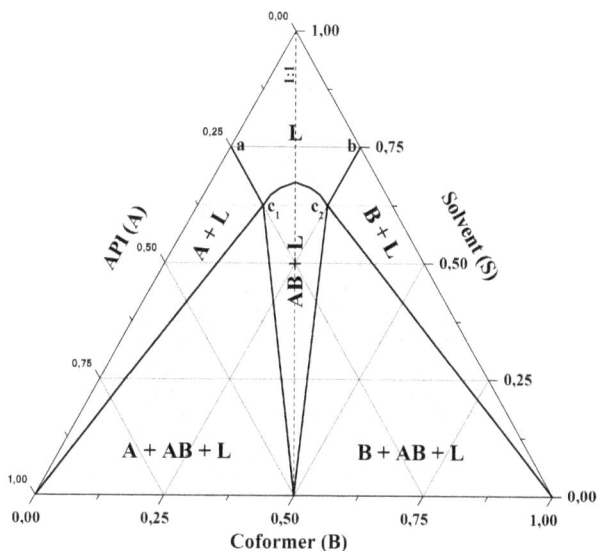

(a)

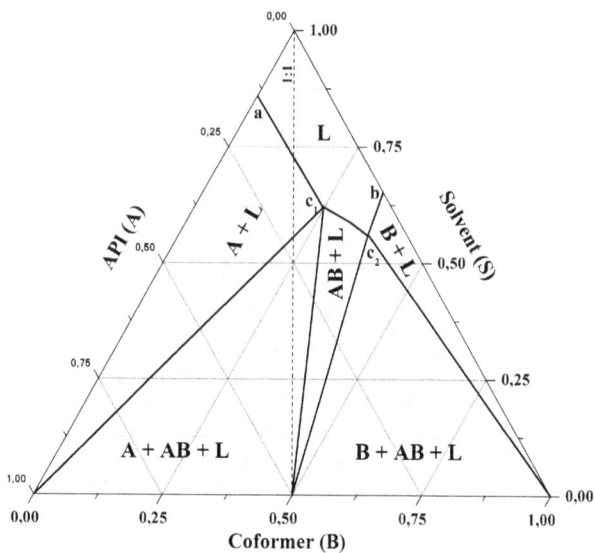

(b)

Figure 5.24. Theoretical TPD for: (a) congruently saturating systems; (b) incongruently saturating systems. (Adapted with permission from Caridi, A., Di Profio, G., Caliandro, R. *et al.* (2012). Selecting the desired solid form by membrane crystallizers: crystals or cocrystals, *Cryst. Growth Des.*, **12**(7), 4349–4356. Copyright 2014 American Chemical Society).

where two solid phases coexist in equilibrium with a liquid phase and define stable thermodynamic regions of the co-crystal (AB) in relation to its pure components. In these systems starting from an equimolar solution (1:1 A:B), the formation of co-crystals occurs. Figure 5.24b is consistent with the more common behavior of a non-congruently saturating system, where an equimolar solution of the two components will not help in producing the pure co-crystal. To obtain the pure co-crystal phase, an initial solution with a non-equimolar composition, and a molar ratio shifted toward the more soluble component, is necessary.[97] In this last case, the necessity of operating in the proper zone of the diagram — avoiding the overrun of the region of a definite species by excessive/uncontrolled supersaturation — in order to obtain selectively the desired product is evident.

The alternative equilibrium phase diagram shown in Fig. 5.25, contains the CBZ solubility curve in the presence of increasing SAC, the SAC solubility curve in the

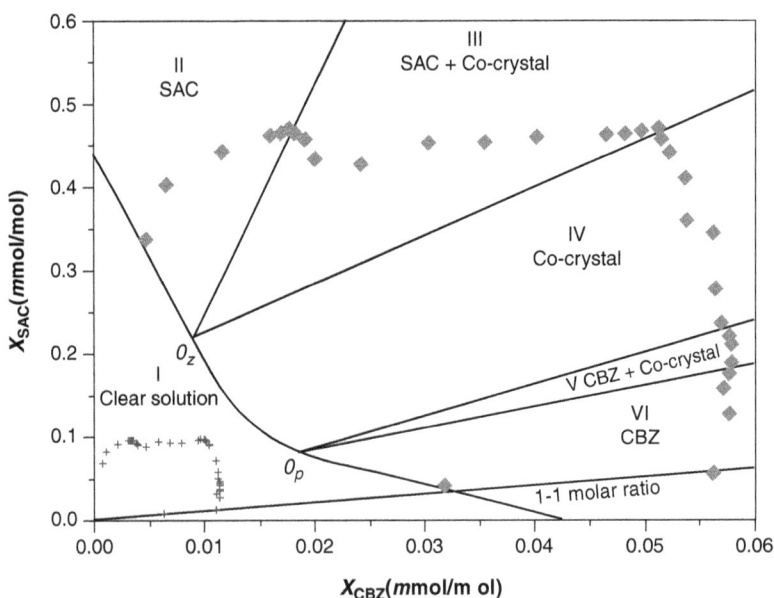

Figure 5.25. Solubility phase diagram of a CBZ-SAC system in 15 wt.%/85 wt.% ethanol/water mixture at 25°C. The determination of solubility curves allows the graph to be approximately divided into six regions: (I) undersaturated clear solution; (II) SAC crystals are stable; (III) mixtures of SAC crystals and CBZ-SAC co-crystal appear; (IV) only CBZ-SAC co-crystals are stable; (V) mixture of CBZ crystals and CBZ-SAC co-crystals exist; (VI) only CBZ crystals are stable. Circles indicate the starting points while diamonds display the crystallization points in membrane crystallization tests. (Adapted with permission from Caridi, A., Di Profio, G., Caliandro, R. *et al.* (2012). Selecting the desired solid form by membrane crystallizers: crystals or cocrystals, *Cryst. Growth Des.*, **12(7)**, 4349–4356. Copyright 2014 American Chemical Society).

presence of increasing amount of CBZ, and the CBZ-SAC co-crystal solubility data in the presence of single components. The curves delimitate the different occurrence regions of the several species and individuate the eutectic points from their intersections.

Once the solubility phase diagram is known for a certain molecular system, the selection of the operating point in any one of the different stability zones allowed directing the crystallization process towards either the preferential formation of the CBZ-SAC co-crystals or the CBZ and SAC single components crystals, by simply choosing the starting conditions. The selection of the proper CBZ/SAC initial molar ratio and the limitation of the supersaturation level to the proper zone of the phase diagram by solvent removal through the porous membrane is the basic requisite to select the specific product.

Figure 5.25 shows a trend consistent with the behavior of the non-congruently saturating systems,[97] where an equimolar solution of the two components will not provide supersaturation specifically for the co-crystal. An initial solution with a non-equimolar composition, where the molar ratio is shifted towards the more soluble component, is necessary to get co-crystals.[108] Indeed, by changing the CBZ/SAC molar ratio in favor of SAC (the more soluble component) the working point migrates to a region of the phase diagram where the co-crystal is the only thermodynamically stable solid phase. Therefore, the choice of suited initial solution compositions (circles in Fig. 5.25) enables the creation of specific pathways in the phase diagram, leading to crystallization points (diamonds in Fig. 5.25) and giving rise to the selective formation of the desired product with the required purity. However, in order to achieve the selective formation of the desired product, the occurrence region of that specific product cannot be overcome. To achieve this purpose, the role of the membrane is crucial to control the degree of solution concentration, thus limiting the maximum level of supersaturation. On these bases, MCr provides the possibility to operate in the proper zone of the phase diagram of the system by controlling the solvent evaporation rate and offers the possibility to design scalable co-crystallization processes.

References

1. Myerson, A.S. and Ginde, R. (1993). in Myerson, A.S. (ed.), *Handbook of Industrial Crystallization*, Butterworth-Heinemann: Boston, p. 33.
2. Volmer, M. and Weber, A. (1926). Keimbildung in übersättigten Gebilden, *Z. Phys. Chem.*, **119**, 227–301.
3. Lewiner, F., Fevotte, G., Klein, J.P. *et al.* (2002). An online strategy to increase the average crystal size during organic batch cooling crystallization, *Ind. Eng. Chem. Res.*, **41**, 1321–1328.

4. Reutzel-Edens, S.M. (2006). Achieving polymorph selectivity in the crystallization of pharmaceutical solids: Basic considerations and recent advances, *Curr. Op. Drug Disc. Dev.*, **9**, 806–815.

5. Mersmann, A. (ed.) (1980). *Thermische Verfahrenstechnik*, Springer, Berlin, Heidelberg.

6. Ulrich, J. and Strege, C.J. (2002). Some aspects of the importance of metastable zone width and nucleation in industrial crystallizers, *Cryst. Growth*, **237/239**, 2130–2135.

7. York, P. (1983). Solid-state properties of powders in the formulation and processing of solid dosage forms, *Int. J. Pharm.*, **14**, 1–28.

8. Schroder, J. (1984). Morphology of organic pigments with special reference to copper phthalocyanine, *Prog. Org. Coatings*, **12**, 181.

9. Hao, Z. and Iqbal, A. (1997). Some aspects of organic pigments, *Chem. Soc. Rev.*, **26**, 203–213.

10. Byrn, S.R. (1982). *Solid State Chemistry of Drugs*, Academic Press, New York, p. 136.

11. Prasad, K.V., Ristic, R.I., Sheen, D.B. *et al.* (2001). Crystallisation of Paracetamol from Aqueous Solution in the Presence and Absence of the Impurity, *Int. J. Pharm.*, **215**, 29–44.

12. Yang, G., Kubota, N., Sha, Z. *et al.* (2006). Crystal shape control by manipulating supersaturation in batch cooling crystallization, *Cryst. Growth Des.*, **6**, 2799–2803.

13. Di Profio, G., Tucci, S., Curcio, E. *et al.* (2007). Controlling polymorphism with membrane-based crystallizers: Application to form I and II of paracetamol, *Chem. Mat.*, **19**, 2386–2388.

14. Ristic, R.I., Finnie, S., Sheen, D.B. *et al.* (2001). Macro- and micromorphology of monoclinic paracetamol grown from pure aqueous solution, *J. Phys. Chem. B*, **105**, 9057–9066.

15. McCrone, W.C. (1965). 'Polymorphism' in Fox, D., Labes, M.M. and Weissberger, A. (eds.), *Physics and Chemistry of the Organic Solid State, Vol. 2*, Wiley Interscience, New York, p. 725.

16. Bernstein, J., Davey, R.J., Henck, J.-O. *et al.* (1999). Concomitant polymorphs, *Angew. Chem. Int. Ed.*, **38**, 3440–3461.

17. Bernstein, J. (1987). 'Conformational polymorphism', in Desiraju, G. (ed.), *Organic Solid State Chemistry*, Elsevier, Amsterdam, p. 471.

18. Price, S.L. (2004). The computational prediction of pharmaceutical crystal structures and polymorphism, *Adv. Drug Deliv. Rev.*, **56**, 301–319.

19. Ostwald, W. (1885). *Lehrbuch der allgemeinen. Chemie II*, Engelmann, Leipzig, p. 440.

20. Hulliger, J. (1994). Chemistry and crystal growth, *Angew. Chem. Int. Ed.*, **33**, 143–162.

21. Sato, K. and Boistelle, R. (1984). Stability and occurrence of polymorphic modifications of stearic acid in polar and nonpolar solutions, *J. Cryst. Growth*, **66**, 441–450.

22. Horne, R.S. and Murthy, K.S.K. (2005). Control of the product quality in batch crystallizationof pharmaceuticals and fine chemicals. Part 1: Design of the crystallization-process and the effect of solvent, *Org. Proc. Res. Dev.*, **9**, 858–872.

23. Elliot, S. (1998). *The Physics and Chemistry of Solids*, Wiley, Chichester, p. 37.

24. Ostwald, W. (1897). Studien über die Bildung und Umwandlung fester Körper, *Z. Phys. Chem.*, **22**, 289–330.
25. Threlfall, T. (2003). Structural and Thermodynamic Explanations of Ostwald's Rule, *Org. Proc. Res. Dev.*, **7**, 1017–1027.
26. Ferrari, E.S. and Davey, R. (2004). Solution-mediated transformation of α to β L-glutamic acid: Rate enhancement due to secondary nucleation, *Cryst. Growth Des.*, **4**, 1061–1068.
27. Davey, R.J. and Cardew, P.T. (1986). Rate controlling processes in solvent-mediated phase transformations, *J. Cryst. Growth*, **79**, 648–653.
28. Cardew, P.T. and Davey, R.J. (1985). The kinetics of solvent-mediated phase transformations, *Proc. Roy. Soc. London A*, **398**, 415–428.
29. Yu, L. (2003). Nucleation of one polymorph by another, *J. Am. Chem. Soc.*, **125**, 6380–6381.
30. Becker, R. and Doering, W. (1935). The kinetic treatment of nuclear formation in supersaturated vapors, *Ann. Phys.*, **5**, 719–752.
31. Roberts, K.J., Docherty, R., Bennema, P. *et al.* (1993). The Importance of considering growth-induced conformational change in predicting the morphology of benzophenone, *J. Phys. D: Appl. Phys.*, **26**, B7–B21.
32. Burton, W.K., Carbera, N. and Franks, F.C. (1951). The growth of crystals and the equilibrium structure of their surfaces, *Phil. Trans. Roy. Soc. London*, **243**, 299–358.
33. Kitamura, M. (2002). Controlling factor of polymorphism in crystallization process, *J. Cryst. Growth*, **237/239**, 2205–2214.
34. Cabri, W., Ghett, P., Pozzi, G. *et al.* (2007). Polymorphisms and patent, market, and legal battles: Cefdinir case study, *Org. Process. Res. Dev.*, **11**, 64–72.
35. Knapman, K. (2000). Polymorphic Predictions: Understanding the Nature of Crystalline Compounds Can Be Critical in Drug Development and Manufacture, *Mod. Drug Discovery*, **3**, 53–57.
36. Ferrari, E.S., Davey, R.J., Cross, W.I. *et al.* (2003). Crystallization in polymorphic systems: The solution-mediated transformation beta to alpha glycine, *Cryst. Growth Des.*, **3**, 53–60.
37. Bauer, J., Spanton, S., Henry, R. *et al.* (2001). Ritonavir: An extraordinary example of conformational polymorphism, *J. Pharm. Res.*, **18**, 859–866.
38. Lucas, J. and Burgess, P. (2004). When form equals substance: The value of form screening in product life-cycle management, *Pharma Voice*, February, 54–57.
39. Bernstein, J. (2002). *Polymorphism in Molecular Crystals*, Oxford University Press, Oxford.
40. ICH Steering Committee (2000). 'ICH Harmonized Tripartite Guideline' in *Good Manufacturing Practice Guide For Active Pharmaceutical Ingredients*, Q7a.
41. Morissette, S.L., Almarsson, Ö., Peterson, M.L. *et al.* (2004). High-throughput crystallisation: Polymorphs, salts, co-crystals and solvates of pharmaceutical solids, *Adv. Drug Deliv. Rev. B*, 275–300.
42. Henck, J.O., Griesser, U. and Burger, A. (1997). Polymorphie von Arzneistoffen, *Pharm. Ind.*, **59**, 165–169.
43. Rodríguez-Spong, B., Price, C.P., Jayasankar, A. *et al.* (2004). General principles of pharmaceutical solid polymorphism: A supramolecular perspective, *Adv. Drug Deliv. Rev.*, **56**, 241–274.

44. Kitamura, M. (1995). *Crystal Growth Handbook*, Japanese Association for Crystal Growth, Kyoritsu-Shuppan, p. 545.
45. Addadi, L., Berkovitch-Yellin, Z., Weissbuch, I. *et al.* (1985). Growth and Dissolution of Organic Crystals with "Tailor-Made" Inhibitors — Implications in Stereochemistry and Materials Science, *Angew. Chem. Int. Ed.*, **24**, 466–485.
46. Davey, R.J., Blagden, N., Potts, G.D. *et al.* (1997). Polymorphism in molecular crystals: stabilization of a metastable form by conformational mimicry, *J. Am. Chem. Soc.*, **119**, 1767–1772.
47. Threlfall, T. (2000). Crystallisation of Polymorphs: Thermodynamic Insight into the Role of Solvent, *Org. Proc. Res. Dev.*, **4**, 384–390.
48. Chen, B.-D., Cilliers, J.J., Davey, R.J. *et al.* (1998). Templated nucleation in a dynamic environment: Crystallization in foam lamellae, *J. Am. Chem. Soc.*, **120**, 1625–1626.
49. Kitamura, M. (1989). Polymorphism in the crystallization of L-glutamic acid, *J. Cryst. Growth*, **96**, 541–546.
50. Marsh, R.E. (1958). A refinement of the crystal structure of glycine, *Acta Crystallogr.*, **11**, 654–663.
51. Iitaka, Y. (1960). The crystal structure of β-glycine, *Acta Crystallogr.*, **13**, 35–45.
52. Iitaka, Y. (1961). The crystal structure of γ-glycine, *Acta Crystallogr.*, **14**, 1–10.
53. Dawson, A., Allan, D.R., Belmonte, S.A. *et al.* (2005). Effect of High Pressure on the Crystal Structures of Polymorphs of Glycine, *Cryst. Growth Des.*, **5**, 1415–1427.
54. Roy, J.J. and Abraham, T.E. (2004). Strategies in making cross-linked enzyme crystals, *Chem. Rev.*, **104**, 3705–3722.
55. Boldyreva, E.V., Drebushchak, V.A., Drebushchak, T.N. *et al.* (2003). Polymorphysm of Glycine. Thermodynamic Aspects Part I: Relative. stability of the polymorphs, *J. Therm. Anal. Calorim.*, **73**, 409–418.
56. Weissbuch, I., Leiserowitz, L. and Lahav, M. (1994). "Tailor-Made" and charge-transfer auxiliaries for the control of the crystal polymorphism of glycine, *Adv. Mater.*, **6**, 952–956.
57. Zaccaro, J., Matic, J., Myerson, A.S. *et al.* (2001). Non-photochemical, laser-induced nucleation of supersaturated aqueous glycine produces unexpected γ-polymorph, *Cryst. Growth Des.*, **1**, 5–8.
58. Lee, A.Y., Lee, I.S., Dette, S.S. *et al.* (2005). Crystallization on confined engineered surfaces: A method to control crystal size and generate different polymorphs, *J. Am. Chem. Soc.*, **127**, 14982–14983.
59. Allen, K., Davey, R.J., Ferrari, E. *et al.* (2002). The crystallization of glycine polymorphs from emulsions, microemulsions, and lamellar phases, *Cryst. Growth Des.*, **2**, 523.
60. Di Profio, G., Tucci, S., Curcio, E. *et al.* (2007). Selective glycine polymorph crystallization by using microporous membranes, *Cryst. Growth Des.*, **7**, 526–530.
61. Boldyreva, E.V., Drebushchak, V.A., Drebushchak, T.N. *et al.* (2003). Polymorphism of glycine, part II, *J. Therm. Anal. Calorim.*, **73**, 419–428.
62. Perlovich, G.L., Hansen, L.K. and Brauer-Brandl, A. (2001). The Polymorphism of Glycine Thermochemical and Structural Aspects, *J. Therm. Anal. Calorim.*, **66**, 699–715.
63. Haisa, M., Kashino, S., Kawai, R. *et al.* (1976). The Monoclinic Form of p-Hydroxyacetanilide, *Acta Crystallogr. B*, **32**, 1283–1285.

64. Haisa, M., Kashino, S. and Maeda, H. (1974). The orthorhombic form of p-hydroxyacetanilide, *Acta Crystallogr. B*, **30**, 2510–2512.

65. Burley, J.C., Duer, M.J., Stein, R.S. *et al.* (2007). Enforcing Ostwald's rule of stages: Isolation of paracetamol forms III and II, *Eur. J. Pharm. Sci.*, **31**, 271–276.

66. Fachaux, J.-M., Guyot-Hermann, A.-M., Guyot, J.-C. *et al.* (1995). Pure Paracetamol for Direct Compression 2. Study of the Physicochemical and Mechanical-Properties of Sintered-Like Crystals of Paracetamol, *Powder Technol.*, **82**, 129–132.

67. Di Martino, P., Guyot-Hermann, A.M., Conflant, P. *et al.* (1996). A new pure paracetamol for direct compression: The orthorhombic form, *Int. J. Pharm.*, **128**, 1–8.

68. Beyer, T., Day, G.M. and Price, S.L. (2001). The prediction, morphology, and mechanical properties of the polymorphs of paracetamol, *J. Am. Chem. Soc.*, **123**, 5086–5094.

69. Joiris, E., Di Martino, P., Berneron, C. *et al.* (1998). Compression behavior of orthorhombic paracetamol, *Pharm. Res.*, **15**, 1122–1130.

70. Nichols, G. and Frampton, C.S. (1998). Physicochemical characterization of the orthorhombic polymorph of paracetamol crystallized from solution, *J. Pharm. Sci.*, **87**, 684–693.

71. Mikhailenko, M.A. (2004). Growth of large single crystals of the orthorhombic paracetamol, *J. Cryst. Growth*, **265**, 616–618.

72. Shekunov, B.Y. and Grant, D.J.W. (1997). *In situ* optical interferometric studies of the growth and dissolution behaviour of paracetamol (acetaminophen). 1. Growth kinetics, *J. Phys. Chem. B*, **101**, 3973–3979.

73. Shekunov, B.Y., Aulton, M.E., Adama-Acquah, R. *et al.* (1996). Effect of temperature on crystal growth and crystal properties of paracetamol, *J. Chem. Soc., Faraday Trans.*, **92**, 439–444.

74. Mikhailenko, M.A., Shakhtshneider, T.P., Drebushchak, T.N. *et al.* (2006). Initial Steps of Dissolution of Monoclinic and Rhombic Crystals of Paracetamol, *Russ. J. Gener. Chem.*, **76**, 48–54.

75. Ivanova B.B. (2005). Monoclinic and orthorhombic polymorphs of paracetamol-solid state linear dichroic infrared spectral analysis, *J. Mol. Struct.*, **738**, 233–238.

76. Di Martino, P., Conflant, P., Drache, M. *et al.* (1997). Preparation and physical characterization of forms II and III of paracetamol, *J. Therm. Anal.*, **48**, 447–458.

77. Söhnel, O. and Müllin, J.W. (1988). Interpretation of crystallization induction periods, *J. Colloid. Interf. Sci.*, **123**, 43–50.

78. Lang, M., Grzesiak A.L. and J. Matzger, A.J. (2002). The Use of Polymer Heteronuclei for Crystalline Polymorph Selection, *J. Am. Chem. Soc.*, **124**, 14834–14835.

79. Price, C.P., Grzesiak, A.L. and Matzger, A.J. (2005). Crystalline polymorph selection and discovery with polymer heteronuclei, *J. Am. Chem. Soc.*, **127**, 5512–5517.

80. Chen, J., Sarma, B., Evans, J.M.B. *et al.* (2011). Pharmaceutical Crystallization, *Cryst. Growth Des.*, **11**, 887–895.

81. Basanta, B., Reddy, K. and Karunakar, A. (2011). Biopharmaceutics classification system: a regulatory approach, *Dissolution Technologies*, February, 31–37.

82. Grzesiak, A.L., Lang, M., Kim, K. *et al.* (2003). Comparison of the Four Anhydrous Polymorphs of Carbamazepine and the Crystal Structure of Form I, *J. Pharm. Sci.*, **92**, 2260–2271.

83. Arlin, J.-P., Price, L.S., Price, S.L. *et al.* (2011). A strategy for producing predicted polymorphs: Catemeric carbamazepine form V, *J. Chem. Commun.*, **47**, 7074–7076.

84. Caridi, A., Di Profio, G., Caliandro, R. *et al.* (2012). Selecting the desired solid form by membrane crystallizers: Crystals or cocrystals, *Cryst. Growth Des.*, **12**, 4349–4356.

85. Desiraju, G.R. (2007). Crystal Engineering: A holistic approach, *Angew Chem. Int. Ed.*, **46**, 8342–8356.

86. Das, D., Banerjee, R., Mondal, R. *et al.* (2006). Synthon evolution and unit cell evolution during crystallisation. A study of symmetry-independent molecules $(Z' > 1)$ in crystals of some hydroxy compounds, *Chem. Commun.*, **5**, 555–557.

87. Anderson, K.M. and Steed, J.W. (2007). Comment on "On the presence of multiple molecules in the crystal asymmetric unit $(Z' > 1)$" by Gautam R. Desiraju, CrystEng-Comm, 2007, 9, 91, *Cryst. Eng. Comm.*, **9**, 328–330.

88. Steed, J.W. (2003). Should solid-state molecular packing have to obey the rules of crystallographic symmetry?, *Cryst. Eng. Comm.*, **5**, 169–179.

89. He, G., Bhamidi, V., Wilson, S.R. *et al.* (2006). Direct growth of glycine from neutral aqeuous solutions by slow, evaporation-driven crystallization, *Cryst. Growth Des.*, **6**, 1746–1749.

90. Mèndez del Rìo, J.R. and Rousseau, R.W. (2006). Batch and Tubular-Batch Crystallization of Paracetamol: Crystal Size Distribution and Polymorph Formation, *Cryst. Growth Des.*, **6**, 1407–1414.

91. He, G., Bhamidi, V., Tan, R.B.H. *et al.* (2006). Determination of Critical Supersaturation from Microdroplet Evaporation Experiments, *Cryst. Growth Des.*, **6**, 1175–1180.

92. Kim, K.-J. and Mersmann, A. (2001). Estimation of metastable zone width in different nucleation processes, *Chem. Eng. Sci.*, **56**, 2315–2324.

93. Zhang, G.G.Z., Law, D., Schmitt, E.A. *et al.* (2004). Phase transformation considerations during process development and manufacture of solid oral dosage forms, *Adv. Drug Delivery Rev.*, **56**, 371–390.

94. Wang, F., Wachter, J.A., Antosz, F.J. *et al.* (2000). An investigation of solvent-mediated polymorphic transformation of progesterone using in situ Raman spectroscopy, *Org. Process Res. Dev.*, **4**, 391–395.

95. Mitchell, C.A., Yu, L. and Ward, M.D. (2001). Selective nucleation and discovery of organic polymorphs through epitaxy with single crystal substrates, *J. Am. Chem. Soc.*, **123**, 10830–10839.

96. Childs, S.L., Chyall, L.J., Dunlap, J.T. *et al.* (2004). Crystal Engineering Approach to Forming Cocrystals of Amine Hydrochlorides with Organic Acids. Molecular Complexes of Fluoxetine Hydrochloride with Benzoic, Succinic, and Fumaric Acids, *J. Am. Chem. Soc.*, **126**, 13335–13342.

97. Gagnière, E., Mangin, D., Puel, F. *et al.* (2011). Cocrystal Formation in Solution: Inducing Phase Transition by Manipulating the Amount of Cocrystallizing Agent, *J. Cryst. Growth*, **316**, 118–125.

98. Trask, A.V., Motherwell, W.D.S. and Jones, W. (2005). Pharmaceutical Cocrystallisation: Engineering a Remedy for Caffeine Hydration, *Cryst. Growth Des.*, **5**, 1013–1021.

99. Sun, C.C. and Hou, H. (2008). Improving mechanical properties of caffeine and methyl gallate crystals by cocrystallization, *Cryst. Growth Des.*, **8**, 1575–1579.

100. Jones, W., Motherwell, W.D.S. and Trask, A.V. (2006). Pharmaceutical Cocrystals: An Emerging Approach to Physical Property Enhancement, *MRS Bull.*, **31**, 875–879.

101. Bethune, S.J., Schultheiss, N. and Henk, J.-O. (2011). Improving The Poor Aqueous Solubility of Nutraceutical Compound Pterostilbene Through Cocrystal Formation, *Cryst. Growth Des.*, **11**, 2817–2823.
102. Almarsson, O. and Zaworotko, M. (2004). Crystal engineering of the composition of pharmaceutical phases. Do pharmaceutical co-crystals represent a new path to improved medicines?, *J. Chem. Comm.*, **17**, 1889–1896.
103. Sheikh, A.Y., Rahim, S.A., Hammond, R.B. *et al.* (2009). Scalable solution cocrystallization: case of carbamazepine-nicotinamide I, *Cryst. Eng. Comm.*, **11**, 501–509.
104. Leung, D.H., Lohani, S., Ball, R.G. *et al.* (2012). Two Novel Pharmaceutical Cocrystals of a Development Compound — Screening, Scale-up, and Characterization, *Cryst. Growth Des.*, **12**, 1254–1262.
105. Urbanus, J., Roelands, C.P., Verdoes, D. *et al.* (2010). Co-Crystallization as a Separation Technology: Controlling Product Concentrations by Co-Crystals, *Cryst. Growth Des.*, **10**, 1171–1179.
106. Jayasankar, A., Reddy, L.S., Bethune, S.J. *et al.* (2009). Role of cocrystal and solution chemistry on the formation and stability of cocrystals with different stoichiometry, *Cryst. Growth Des.*, **9**, 889–897.
107. Childs, S.L., Rodrìguez-Hornedo, N., Reddy, L.S. *et al.* (2008). Screening strategies based on solubility and solution composition generate pharmaceutically acceptable cocrystals of carbamazepine, *Cryst. Eng. Comm.*, **10**, 856–864.
108. Childs, S.L., Wood, P.A., Rodríguez-Hornedo, N. *et al.* (2009). Analysis of 50 crystal structures containing carbamazepine using the Materials Module of Mercury CSD, *Cryst. Growth Des.*, **9**, 1869–1888.

Chapter 6

Crystallization of Biomacromolecules

6.1 Introduction

The crystallization of biomacromolecules by using membrane-assisted crystal-lization technology is described in this chapter. The discussion will focus on the interest in growing protein crystals and the current difficulties in obtaining them by conventional crystallization methods, the possibility to use MAC for the growth of protein crystals and the influence of controlling the transmembrane flux on crystallization kinetics, the function of polymeric membranes as solid support for heterogeneous nucleation and the correlation between membrane physical prop-erties and the energetics of nucleation, and the effect of forced solution-flow con-vection on crystallization kinetics and its influence on crystal properties.

6.2 Interest in Protein Crystallization

The functions of biomolecules are strictly connected to their three-dimensional molecular structures. Therefore, the determination of the spatial arrangement of atoms or groups of atoms in protein molecules is of fundamental importance for the understanding of their biological activity and for designing new drug molecules that can act at active sites.[1] The 3D structure of proteins can be determined by X-ray or neutron diffraction and nuclear magnetic resonance (NMR) spectroscopy. However, the crystallographic method is the one of choice, especially for molecules that are too large (M.W. > 20 kDa). Crystallography gives a higher resolution, reaching less than 1 Å for particularly complex systems like membrane-protein or viruses.[2] The basic requisite for structure resolution at the atomic level by X-ray diffraction is the availability of well-diffracting crystals of adequate size.[3]

Among proteins, enzymes are the most efficient known catalysts because of their high substrate specificity.[4] However, their systematic utilization in solution at industrial scale has been rather limited due to their intrinsic labile nature under operative conditions. Poor chemical and mechanical stability at extreme temperatures and pH, high pressure, mechanical stress, and the presence of denaturing solvents and other chemicals, are the factors that have prevented enzymes from being extensively used. Also, in the solid state, proteins are normally characterized by pronounced fragility or shattering under relatively mild conditions. On the other hand, the possibility of producing a large amount of protein crystal with reduced, uniform, and predictable size, slow dissolution kinetics, and high chemical and mechanical stability would have remarkable implications in many biotechnological applications, e.g. for sustained constant rate drug release, in environmental protection applications, in chemical synthesis, etc.[5] In this respect, cross-linked enzyme crystals (CLECs)[6,7] are interesting systems whose production is a relatively simple, fast, and less expensive technique and one that does not require chemical stabilizers, extreme temperature, or drying-out steps,[5,8–10] In spite of their increased stability, CLECs' performance depends on the structural and morphological characteristics of the crystalline particles, like polymorphism (when more than one polymorph can be grown), shape, size, and size distribution. These attributes have a profound effect on their physical-chemical parameters, determining their efficacy, stability, melting point, dissolution rate, and so on. In each case, it is well known that only crystals with uniform shape and size display high catalytic efficiency.[5]

According to the above discussion, the interest in biomacromolecules crystallization in several scientific and industrial sectors demands technologies that can grow these molecules under conditions that generate crystals with selected morphological and structural properties.

6.3 Difficulties in Obtaining Protein Crystals

Although, macromolecular crystals grow from solutions by the same mechanisms as small molecules, they show some peculiarities that make crystallization much more challenging:

(1) Macromolecules display a rather asymmetric and weak bonding configuration at their surfaces, thus making ordered attachment for growing units less probable;

(2) They tend to aggregate in n-mers that diversify the shape and size of the lattice units, making the ordered growth more complex;

(3) Solutions typically contain considerable amounts of contaminants and impurities, whose incorporation in the crystal structures represents an additional difficulty to their growth;

(4) Usually two or more solid states can appear in the system – not only crystals, but also precipitates, oils, or gels;

(5) Macromolecules are complex systems which display very low diffusivities, so that they show too-slow crystallization kinetics with respect to small molecules.

As a result, protein crystallization is a complicated physical-chemical process: a specific macromolecule often crystallizes in a narrow range of several parameters, including concentration, temperature, pH, type and concentration of precipitating agent, presence of counter-ions, impurities, and other chemical additives, as well as method and rate of supersaturation generation. In addition, protein crystallization from solution presents added complexities stemming from the intrinsic evolutionary development towards the avoidance of spontaneous aggregation in the living cell environment.[11] This means that different methods of crystallizing the same protein may give rise to different quantities and qualities of crystals even under the same state of the mother liquor.[3]

The most frequently used protein crystallization method is vapor diffusion. In this technique, the protein is mixed with an equal volume of the crystallizing solution (usually buffer, salt, and precipitant) and equilibrated with a reservoir solution by diffusion in the vapor phase. Three different setups exist (Fig. 6.1):

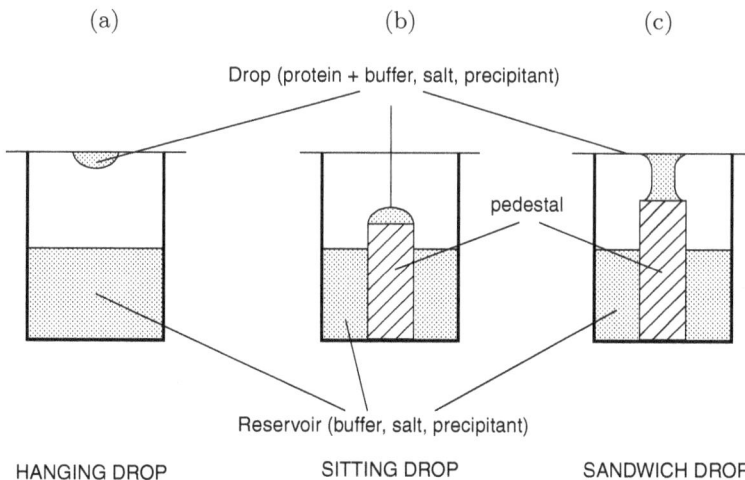

Figure 6.1.   The three types of vapor diffusion setups: (a) hanging drop, (b) sitting drop, (c) sandwich drop.

hanging drop, sitting drop, and sandwich drop. In batch technique, the precipitant and the target molecule solution are simply mixed, and supersaturation is achieved directly. Although very useful in crystallography, the vapor diffusion methodology is limited to the production of small amount of crystals for X-ray analysis, while the batch is a simple and low cost technique, currently used in industry but characterized by poor control, which might lead to unpredictable crystals features.

6.4 Regulation of Supersaturation Rate in Membrane-Assisted Crystallization of Proteins

Nucleation and crystal growth rates are supersaturation-dependent. Usually, low supersaturation favors crystal growth while higher values are associated with excessive nucleation. The consequence of the supersaturation degree in a crystallization trial is the relative extent of the nucleation over the crystal growth rate, while the macroscopic evidence is on crystal morphology and size.

In membrane-assisted crystallization (MAC), the membrane acts as a dosing device to generate supersaturation. Mass transfer, quantified in terms of transmembrane flux J, is fixed by the proper combination of the operational parameters and membrane properties, which affect the specific driving force. In the case of membrane distillation and osmotic distillation-based MAC, according to the dusty gas model equation simplified for the Knudsen-limited diffusion through a porous material, the transmembrane flux is given by[12]:

$$J = \frac{2\varepsilon r}{3\tau} \frac{1}{RT} \left(\frac{8RT}{\pi M}\right)^{\frac{1}{2}} \frac{\Delta p}{\delta}, \tag{6.1}$$

where R is the gas constant, T the temperature, M the molecular weight, Δp the gradient of partial pressure, ε is the porosity, τ the tortuosity, r the pore size, and δ the thickness of the membrane.

In protein crystallization, the macromolecule, precipitant agent, buffer, and potential additives, compose the crystallizing solutions. Several parameters like solute, precipitant, and draw solution concentration, have influence on the flux by acting on the partial pressure of the solvent (water) in the crystallizing solution. Generally, increased concentration of components in solution is associated to a reduced driving force for solvent evaporation. Additionally, as crystallizing solution concentration increases because of solvent evaporation, the driving force decreases even more, thus reducing the rate of solution concentration (transmembrane flux). To maintain a constant transmembrane flux and supersaturation rate, changes in solution composition because of solvent removal can be counterbalanced by corrections in the driving force by acting on operative temperature and stripping solution concentration. These parameters can be controlled in time during the crystallization as the macromolecular solution increases its concentration, so that the state of the system moves along a well-defined trajectory in the phase diagram. Therefore, the extent of the crystalline population, the morphological (size, habit, and shape) and structural (polymorphism) crystals properties, as well as purity can be controlled.

Supersaturation is thermodynamically defined by the characteristic solubility of the solute. Protein solubility is a complex property that is sometimes difficult to predict. Under typical conditions, proteins are charged macromolecules with long-range interactions, and deviations from the ideality of their solutions is taken into account by means of the activity coefficient γ, commonly estimated from measurements of the second virial coefficient B_{22}. This coefficient derives from the virial expansion for the activity coefficient as function of concentration c:

$$\ln \gamma = 2B_{22}c + \frac{3}{2}B_{33}c^2 + \cdots \approx 2B_{22}c. \tag{6.2}$$

Here, the standard state is that $\gamma \rightarrow 1$ as $c \rightarrow 0$. In dilute solutions, binary interactions prevail over ternary interactions, so it usually suffices to keep just the second virial coefficient for interactions between pairs of particles.[13] This quantity provides a measure of the dilute protein-protein interactions that are Boltzmann-averaged over all distances and orientations in the liquid phase.[14] The equilibrium solubility of globular proteins correlates with the thermodynamic parameter B_{22},[15] that is an integral measure of the strength of protein–protein interactions, the mean-force potential.[14] Accordingly, B_{22} can also be defined as

$$B_{22} = 2\pi \int_0^\infty (1 - e^{-U(r)/k_B T}) \cdot r^2 dr. \tag{6.3}$$

where r is the center-to-center distance of the particles, k_B is the Boltzmann constant, T is the absolute temperature and $U(r)$ is the pair interaction potential. In the equation it does not matter what is added to the protein solution to achieve a given value of B_{22}; once the strength of attraction has been achieved, the protein will have the same solubility. Therefore, evaluation of B_{22} can give a qualitative measure of the protein-protein interactions in a dilute solution: if B_{22} is positive, the net interaction between protein molecules is repulsive; if B_{22} is negative, the net interaction between protein molecules is attractive.

A necessary, but insufficient, requirement for the success of protein crystallization is that values of B_{22} vary in a precise range (crystallization slot).[16] It is also known that these values of B_{22} correspond to protein-protein interactions that are short-ranged with respect to the size of the protein molecule. This condition typically corresponds to salt concentrations sufficiently large to screen the Coulomb repulsive interaction, so that the Debye screening length is small.[17] For lysozyme, B_{22} should be in the interval $-2 \times 10^{-4}\,\text{mol·mL·g}^{-2}$ to $-8 \times 10^{-4}\,\text{mol·mL·g}^{-2}$. Progression towards more negative values corresponds to an increase of protein-protein interactions. For B_{22} greater than $-2 \times 10^{-4}\,\text{mol·mL·g}^{-2}$, the intermolecular attraction is usually not sufficiently strong to form stable crystals. A B_{22} lower than $-8 \times 10^{-4}\,\text{mol·mL·g}^{-2}$ implies that the protein-protein attractions are so strong that the protein molecules may not have adequate time to orient themselves to form a crystalline lattice, sometimes leading to an amorphous precipitate. Positive values of B_{22} generally indicate larger protein-solvent interactions compared to protein-protein interactions, i.e. the net protein intermolecular forces are repulsive. The crystallization slot, in terms of the osmotic second virial coefficient, could point to solution conditions that lead to "good" nucleation. Therefore, B_{22} has a predictive character regarding protein crystallization and/or precipitation from solution.

It follows from above that the osmotic second virial coefficient correlates with nucleation kinetics[18] as nucleation kinetics correlate with the protein solubility. As B_{22} is determined in a non-destructive manner from dilute solutions, it forms a useful guideline while attempting the crystallization of proteins by manipulating solution conditions to a place where proteins will readily crystallize. The selection of solution composition thus requires a delicate balance to provide the proper value of B_{22} conducive to crystallization. In this respect, MAC provides the possibility to adjust the solution concentration acting on the transmembrane flux, thus affecting the activity coefficient and, as a consequence, B_{22}. In general, an increase in both precipitant and protein concentrations results in the reduction of the activity of the mother liquor, more markedly at higher content of precipitant.

Figures 6.2 and 6.3 show the influence of process parameters such as the concentration of precipitating agent (NaCl) and draw solution ($MgCl_2$) on water flux through the membrane in the crystallization of lysozyme.

For high concentrations of NaCl, the second virial coefficient systematically progresses towards more negative values, corresponding to increasing protein-protein attractions.[15] An increase in NaCl from 0.75% to 2% w/v at 20°C determines a decrease in B_{22} from $-0.5 \times 10^{-4}\,\text{mol·mL·g}^{-2}$ to $-2.0 \times 10^{-4}\,\text{mol·mL·g}^{-2}$, which reflects a transmembrane flux reduction of about 4% (from $0.24\,\mu\text{L/mm}^2\text{·h}$ to $0.23\,\mu\text{L/mm}^2\text{·h}$) for a lysozyme concentration around 30 mg/mL. This depletion is more relevant when the precipitant concentration is further heightened, reaching 20% for a content of 4.5% w/v in NaCl. The reduction of the transmembrane flux consequent to an increase in protein concentration is due to the explicit correlation between this parameter and the activity coefficient. Values for solvent flux of $0.21\,\mu\text{L/mm}^2\text{·h}$ and $1.9\,\mu\text{L/mm}^2\text{·h}$ were measured with

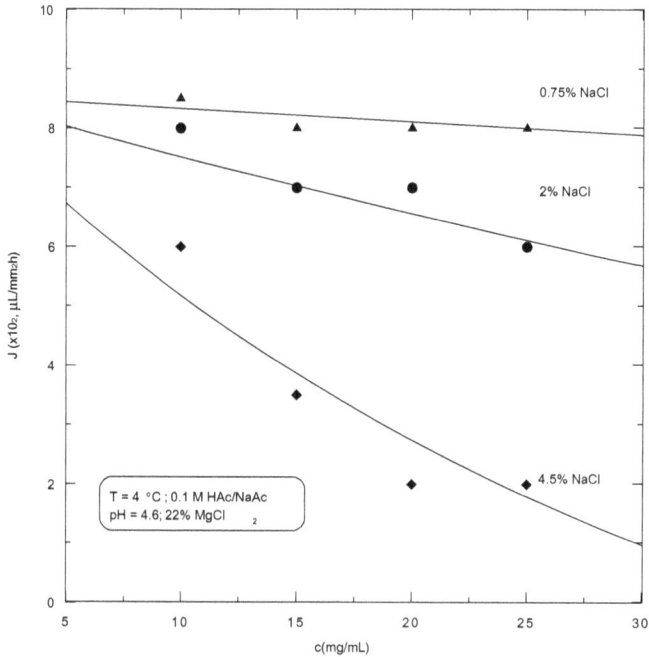

Figure 6.2. Influence of the precipitant content (0.75%–4.5% NaCl) on transmembrane flux and different lysozyme concentrations (5 mg/mL–50 mg/mL) at 20°C. Theoretical (lines) (Eq. (6.3)) and experimental (symbols) data obtained at 22% MgCl$_2$ and pH $= 4.6$. (Reprinted from Curcio, E., Di Profio, G. and Drioli, E. (2003). A new membrane-based crystallization technique: tests on lysozyme, *J. Cryst. Growth*, 247, 166–176. Copyright 2014, with permission from Elsevier).

two solutions prepared by using the same precipitant concentration (4.5% NaCl) but different lysozyme concentrations: 20 mg/mL and 30 mg/mL, respectively.

The effect of temperature and stripping solution on the transmembrane flux between 5°C and 25°C with steps of 5°C, and increasing the concentration of MgCl$_2$ solution from 10% to 22% w/w is shown in Fig. 6.3. Two concomitant effects determine the exponential trend of the flux rate consequent to an increase in temperature. On one hand, the Clausius–Clapeyron relation establishes that the process driving force — a vapor pressure gradient between both membrane interfaces — is an exponential function of the temperature; for example, the vapor pressure of the pure solvent (water) changes from 786 Pa to 2295 Pa when the temperature rises from 4°C to 20°C. On the other, increases in temperature shift the second virial coefficient towards less negative values and this effect is more pronounced at low ionic strength. For lysozyme in 2% w/v NaCl at 20°C, the B_{22} value is -2×10^{-4} mol·mL·g^{-2}; lowering the temperature to 5°C, B_{22} results

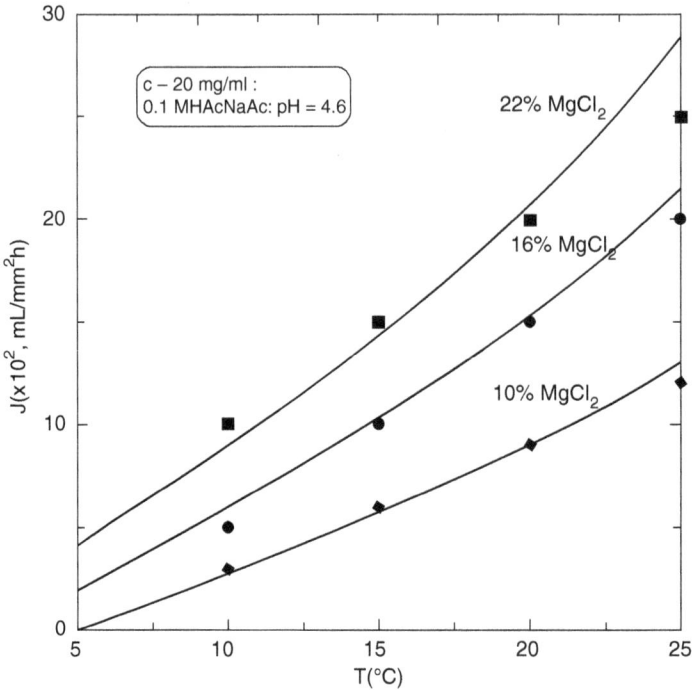

Figure 6.3. Transmembrane flux vs temperature (5°C–25°C) at different stripping concentrations (10%–22% $MgCl_2$). Theoretical (lines) and experimental (symbols) results obtained at constant protein concentration (20 mg/mL), 4.5% NaCl and pH = 4.6. (Reprinted from Curcio, E., Di Profio, G. and Drioli, E. (2003). A new membrane-based crystallization technique: tests on lysozyme, *J. Cryst. Growth*, 247, 166–176. Copyright 2014, with permission from Elsevier).

-5×10^{-4} mol·mL·g^{-2}, thus leading to an increase in the activity coefficient of the solution. In this case, a faster extraction of the solvent due to higher transmembrane flux is counterbalanced by the increase in solubility of the lysozyme from 5 mg/mL at 5°C to 28 mg/mL at 20°C,[15] which involves a greater volume of solvent being removed before to reach supersaturation.

Tuning of the transmembrane flux is easily obtained by adjusting the concentration of the $MgCl_2$; because of the depletion of the activity on the stripping side, faster evaporation rates at constant temperature are found, corresponding to an increase of the salt content, as shown in Fig. 6.4. At 20°C and under conditions reported in the legend, the solvent flux leaps from 0.09 μL/mm^2 to 0.16 μL/mm^2 h when the $MgCl_2$ concentration increases from 10% to 16% w/v. In the case of more concentrated protein solutions, the progressive decline of the system performance is illustrated in Fig. 6.4.

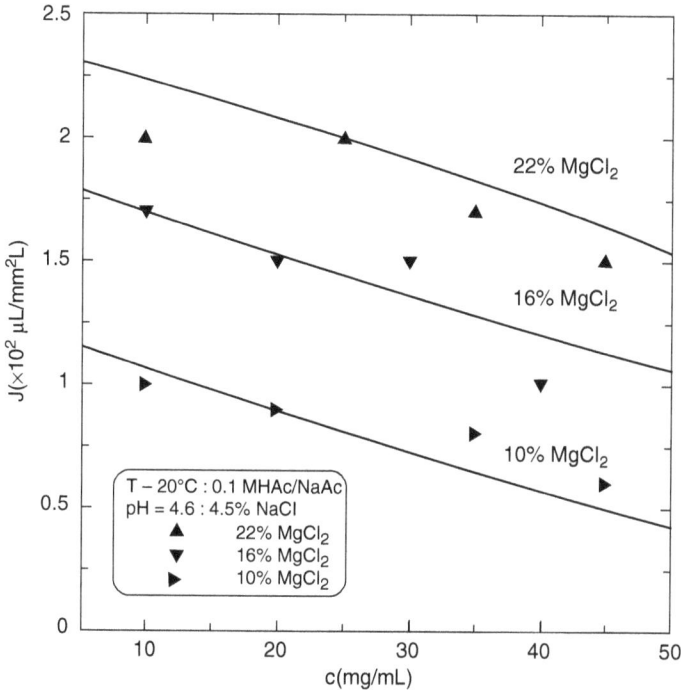

Figure 6.4. Influence of the stripping content (10%–22% $MgCl_2$) on transmembrane flux at different lysozyme concentrations (5 mg/mL–50 mg/mL). Theoretical (lines) and experimental (symbols) data obtained at 4.5% NaCl and pH $= 4.6$. (Reprinted from Curcio, E., Di Profio, G. and Drioli, E. (2003). A new membrane-based crystallization technique: tests on lysozyme, *J. Cryst. Growth*, **247**, 166–176. Copyright 2014, with permission from Elsevier).

6.5 Influence of Supersaturation Control on Crystallization Kinetics

As described in the previous section, two main factors that influence the crystallization process in MAC are the solvent activity in the protein/precipitant and stripping, with the resulting effect on the partial pressure gradient between the two sides of the membrane (Eq. (6.1)) that represents the driving force of the mass transfer, and the effect of the ionic strength on the protein solubility. Therefore, the interactions between these two contributions plays a major role in MAC processes.[19]

Volmer's theory[20] establishes a linear dependence between the logarithm of the nucleation rate and the square of the supersaturation. Therefore, the rate of solvent removal in vapor phase through a porous membrane is expected to affect the crystallization kinetics. As supersaturation control is directly related to

Table 6.1. Different kinds of PVDF membranes: mean pores size, porosity, and thickness.

Code	Mean pore size (μm)	Porosity (−)	Thickness (μm)
PK	0.11	4.6×10^{-1}	87.4
PK-L2	0.14	5.4×10^{-1}	98.9
PK-L5	0.043	4.4×10^{-1}	103
PK-L7	0.056	2.7×10^{-1}	101
PK-P2	0.21	4.0×10^{-1}	122
PK-P5	0.20	4.5×10^{-1}	150
PK-P7	0.22	5.8×10^{-1}	150
PKF	0.034	6.0×10^{-3}	41.4
PKF-L2	0.072	4.3×10^{-2}	69.4
PKF-L5	0.084	5.9×10^{-1}	117
PKF-L7	0.043	3.2×10^{-1}	91.5
PKF-P2	0.044	1.1×10^{-1}	103
PKF-P5	0.092	5.4×10^{-1}	121
PKF-P7	0.19	2.7×10^{-1}	153

transmembrane flux J, physical membrane properties, such as mean pore size, overall porosity, and thickness, which also contribute to J, have influence on crystallization kinetics. Table 6.1 includes some poly(vinylidene) fluoride (PVDF) membranes with different physical properties. The overall effect of the characteristics of the membrane on the crystallization kinetics of proteins is described in the following section.

6.5.1 *Monitoring nucleation and crystal growth rates by the turbidity method*

Turbidity measurements are an easy and immediate method applied to protein crystallization as a diagnostic tool to monitor nucleation and crystal growth[21] and to measure induction time periods.[22] This technique is practically insensitive to multiple scattering[23] and it is suitable for the study of concentrated solutions. The turbidity profile curves showed in Fig. 6.5 show the time interval (induction period) that precedes the formation of stable crystal nuclei, where almost no changes in turbidity can be observed. The simultaneous nucleation and growth of protein clusters resulted in a rapid increase in the turbidity followed by a leveling off. A rapid decrease in turbidity is then observed due to crystal sedimentation. The in-turn repetition of the nucleation, growth, and sedimentation steps can be detected by monitoring the trend in the turbidity peaks.

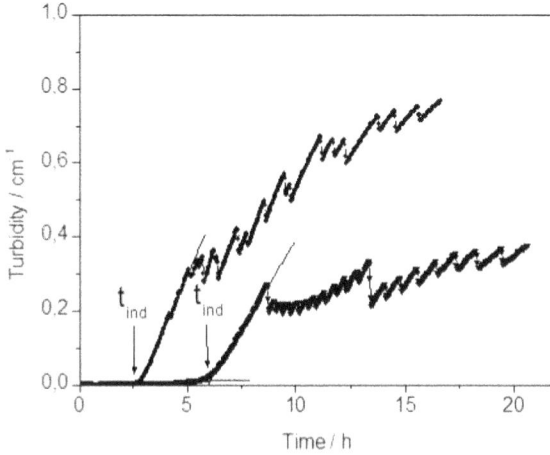

Figure 6.5. Typical turbidity profiles obtained during MAC experiments of lysozyme (HEWL 20 mg/mL in AcNa/AcH buffer 0.1 M, pH = 4.6). (Reprinted from Di Profio, G., Curcio, E., Cassetta, A. *et al.* (2003). Membrane crystallization of lysozyme: kinetic aspects, *J. Cryst. Growth*, 257, 359–369. Copyright 2014, with permission from Elsevier).

The turbidity profile curves in combination with the Rayleigh scattering theory[24] can be used to estimate nucleation and crystal growth rates. For a scattering path length d, the relationship between turbidity τ and absorbance A of a suspension of particles is given by

$$\tau = \frac{2.303}{d} \cdot A. \tag{6.4}$$

According to the Rayleigh scattering theory, for particles having a radius r smaller than the light wavelength λ in the dispersing medium ($r < 0.05\lambda$), the size distribution of poly-dispersed particles is related to the turbidity profile of the solution as follows:

$$\tau = \frac{128\,\pi^5}{3}\frac{1}{\lambda^4}\left(\frac{n^2-1}{n^2+1}\right)^2 \int_0^\infty f(r) \cdot r^6 dr, \tag{6.5}$$

where $f(r) \cdot dr$ represents the fraction of particles with radius r and $r + dr$; n is the ratio between the refractive index of the particles n_p and the refractive index of the medium n_m. Integrating Eq. (6.5) gives the quantification of particles with selected size nucleated in the system.

Nucleation rate B can be estimated as the variation of the number of generated crystals ΔN during the time interval Δt for a given volume unit V:

$$B = \frac{1}{V}\frac{\Delta N}{\Delta t}. \tag{6.6}$$

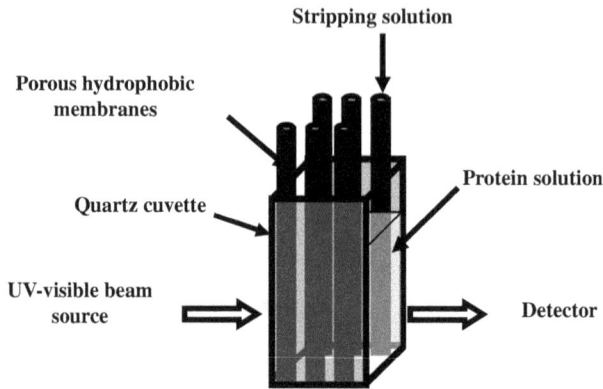

Figure 6.6. A schematic representation of a quiescent membrane system used to monitor *in situ* the turbidity profile of a crystallizing solution. Porous polymeric fibers, filled with MgCl$_2$ solution as a stripping agent, are inserted in a quartz cuvette where the protein/precipitant solution is loaded. (Reprinted from Di Profio, G., Curcio, E., Cassetta, A. *et al.* (2003). Membrane crystallization of lysozyme: kinetic aspects, *J. Cryst. Growth*, 257, 359–369. Copyright 2014, with permission from Elsevier).

Time-evolution of the equivalent radius for a growing particle is assumed to obey to the exponential relationship

$$r(t) = r_{\max} - (r_{\max} - r_0)e^{-k(t-t_0)}, \qquad (6.7)$$

where r_0 is the cluster radius at the instant t_0, r_{max} the final equivalent radius of the particle, and k is the rate constant. Eqs. (6.4)–(6.7) allow assessing the early-stage nucleation kinetic and the crystal growth process. Accordingly, Eq. (6.7) provides a way to estimate particle size in the time.

Figure 6.6 displays a quiescent MAC apparatus for on-line turbidity analysis. It consists of hydrophobic porous hollow fibre membranes arranged in a quartz cuvette. The stripping solution is inside the membrane fibers, whereas the protein/precipitant solution mixture is loaded on the outer part of the membranes. The cuvette is placed in a UV-visible spectrophotometer while absorbance is recorded at 400 nm during that time.

6.5.2 *Kinetic parameters*

In the previous section, it was shown that precipitant and stripping solution concentrations influence transmembrane flux. This effect has a direct consequence in the crystallization kinetics of proteins in MAC, as displayed in the case of hen egg white lysozyme (HEWL) crystallization with NaCl as precipitant and NaCl or MgCl$_2$ as draw solution.

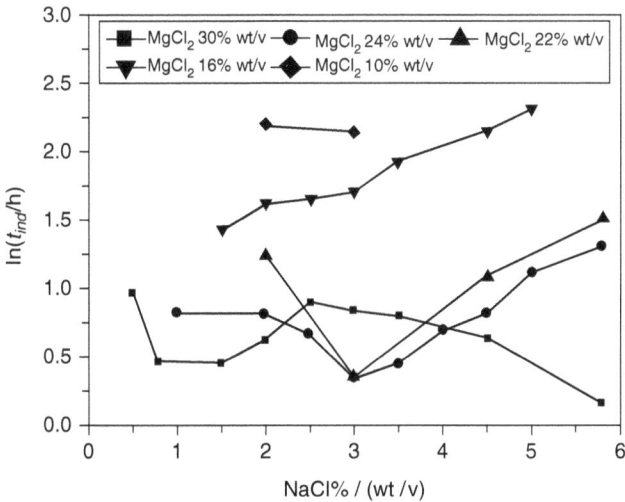

Figure 6.7. Induction times vs. NaCl concentration; MgCl$_2$ % w/v is indicated in the legend. (Reprinted from Di Profio, G., Curcio, E., Cassetta, A. *et al.* (2003). Membrane crystallization of lysozyme: kinetic aspects, *J. Cryst. Growth*, 257, 359–369. Copyright 2014, with permission from Elsevier.)

At a constant NaCl concentration, a generalized decrease in the induction time period is observed when the MgCl$_2$ concentration increases (Fig. 6.7). This is due to the higher rate for solvent extraction, so that solution quickly reaches the required supersaturation level. In the case of constant stripping-solution concentration and variable precipitant composition, a trade-off is observed, because of the kinetic-thermodynamic competition in the solution concentration process. For higher stripping solution concentration (> 16% w/v), the induction time is initially reduced as the content of NaCl is increased. This effect arises because of protein solubility dropping. If the concentration of the precipitant is increased further, the gradient of solvent activity between the stripping and the protein/precipitant solutions falls, thus reducing the driving force for solvent evaporation; this leads to a substantial increase in the induction time. For lower concentration of stripping solution (16% w/v), induction time increases constantly with increasing NaCl concentration; in this case, the dominant factor is the drop in driving force over the precipitating effect.

The variation of nucleation rate at different concentrations of NaCl, used both as precipitant in the protein solution and as stripper on the opposite side of the membrane, is reported in Fig. 6.8 for various PVDF membranes. At a precipitant concentration of 1.5% w/v and 5°C, B_{22} is -3.5×10^{-4} mol·mL·g^{-2} and the transmembrane flux reaches its maximum value of 2.0×10^{-1} μL/cm^2·h.

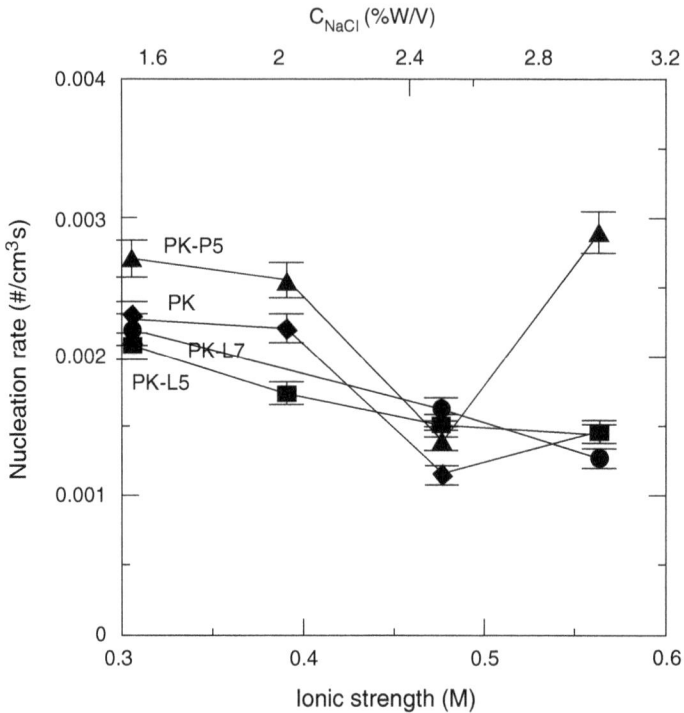

Figure 6.8. HEWL nucleation rates measured on PVDF membranes at 3% w/v NaCl. (Adapted with permission from Curcio, E., Fontananova, E., Di Profio, G. *et al.* (2006). Influence of the structural properties of poly(vinylidene fluoride) membranes on the heterogeneous nucleation rate of protein crystals, *J. Phys. Chem. B*, 110, 12438–12445. Copyright 2014 American Chemical Society).

An increase of NaCl content reduces the activity of the protein solution, thus leading to a progressive depletion of the mass transfer driving force. For NaCl 2.5% w/v, B_{22} continues its progression towards more negative values (-6.2×10^{-4} mol·mL·g^{-2}), the activity coefficient of the crystallizing solution reduces to about 19% and the flux of the vaporized solvent drops off to 1.7×10^{-1} μL/cm^2·h. The experimental condition corresponding to the highest precipitant concentration (3% w/v) is characterized by the lowest value of vapor flux (1.3×10^{-1} μL/cm^2·h). Up to an ionic strength of 0.48 M, the reduced rate of solvent removal affects the supersaturation level of the solution, and the nucleation rate decreases from 2.3×10^{-1} cm^{-3}·s^{-1} to 1.1×10^{-1} cm^{-3}·s^{-1}. In Fig. 6.13, PK and PK-P5 curves exhibit a minimum in the nucleation rate ($\sim 1.5 \times 10^{-1}$ cm^{-3}·s^{-1}) corresponding to 2.5% w/v NaCl. At high precipitant concentrations, the PK-L5 kinetic curve gently approaches its lowest value, while the nucleation rate of lysozyme crystals on the PK-L7 membrane decreases with almost constant slope.

In most cases, as confirmed in Fig. 6.8, the effect of the transmembrane flux on the nucleation rate is counterbalanced by the enhancement of protein–protein interactions at high precipitant concentrations. As a consequence, the saturation of the protein solution decreases, and the crystallization system operates at increased supersaturation levels. At 5°C and 1.5% w/v NaCl, the solubility for lysozyme is about 7 mg/mL, but decreases to <1 mg/mL for precipitant concentration greater than 2.5% w/v. The influence on the crystallization kinetics can be drastic; for instance, as the ionic strength increases from 0.48 M to 0.56 M, the nucleation rate measured on PK-P5 membranes practically doubles, rising from $1.4 \times 10^{-1}\,cm^{-3}\cdot s^{-1}$ to $2.9 \times 10^{-1}\,cm^{-3}\cdot s^{-1}$.

Figure 6.9a illustrates the effect of the NaCl concentration on the growth rate, under the assumption that the crystal size evolves in time according to Eq. (6.7). For a stripping concentration of 30% w/v $MgCl_2$, the initial increase in growth rate constant k due to a NaCl variation from 0.8% w/v to 3% w/v is followed by a considerable deceleration when the NaCl concentration is further raised. This is expected to be determined by the activity gradient between protein/precipitant and stripping solutions (depending on both NaCl and $MgCl_2$ concentrations) and the protein solubility (only influenced by NaCl concentration).

A decline in growth rate has been generally observed when the $MgCl_2$ concentration was increased, as indicated in Figure 6.9b at 4.5% w/v NaCl. On the contrary, k increased when the stripping solution concentration was decreased from 30% w/v to 16% w/v. This trend is in the opposite direction if compared to the nucleation rate; in fact, due to a decrease in $MgCl_2$ concentration, crystals reduced in number but larger in size were produced.

The order of magnitude of k measured in MAC is comparable with data obtained by fitting Eq. (6.7) on a number of experimental size versus time curves reported in the literature[25–27]; higher values of growth rate constant were observed at lower supersaturation ratios than those indicated in the literature.[21,25–27] Maximum values of growth rate obtained using MAC are higher than those found in the literature; this can be explained by the heterogeneous contribution to crystallization provided by the membrane surface.

6.6 Heterogeneous Nucleation by Porous Membranes

Approaches aimed at promoting heterogeneous nucleation provide interesting results in terms of shorter induction times and accelerated nucleation rate of protein

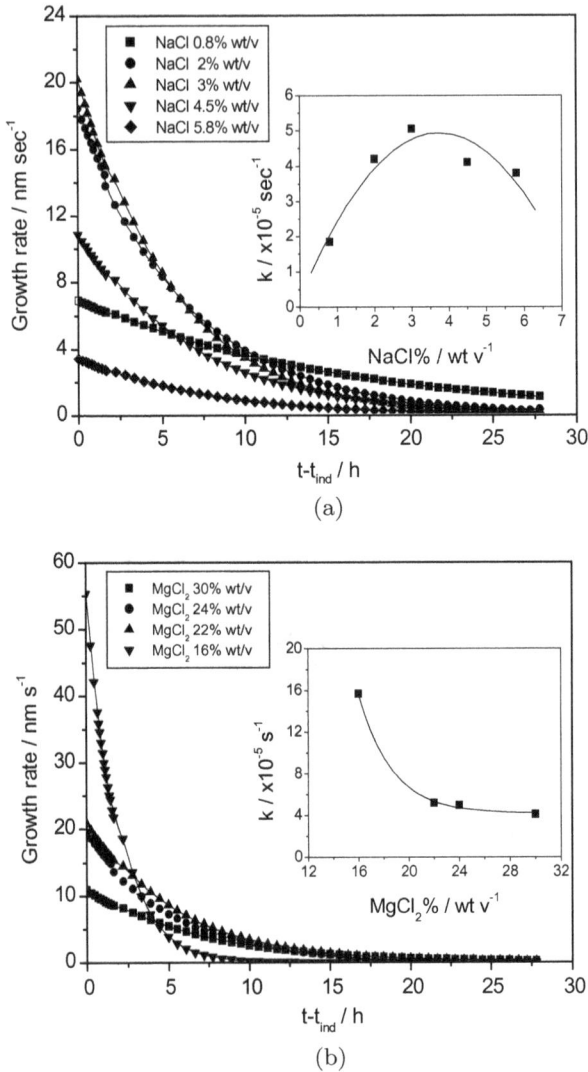

Figure 6.9. Lysozyme growth rates vs. time: (a) 30% w/v MgCl$_2$, NaCl concentrations reported in the legend; (b) 4.5% w/v NaCl, MgCl$_2$ concentrations reported in the legend. In the insets are shown the relative variations of the growth rate constant k. (Reprinted from Di Profio, G., Curcio, E., Cassetta, A. *et al.* (2003). Membrane crystallization of lysozyme: kinetic aspects, *J. Cryst. Growth*, 257, 359–369. Copyright 2014, with permission from Elsevier).

crystals, with respect to those reported by using traditional methods,[28] as well as facilitated crystallization of macromolecules that are reluctant to crystallize. This is the reason why the effect of several substrates on protein nucleation rate has been the object of extensive research. Mineral substrates (including apophyllite, topaz, lepidolite, magnetite, and galena) have been used as support in lysozyme nucleation rate.[29,30] Polymeric films containing ionizable groups, such as sulfonated polystyrene, cross-linked gelatin films with adsorbed poly-L-lysine or entrapped poly-L-aspartate, and silk fibroin with entrapped poly-L-lysine or poly-L-aspartate, have been investigated.[31,32] Crystallization tests carried out on silicon substrates,[33] polymeric films with ionizable groups,[31] glass surfaces made hydrophobic by coating with dimethyl-disilazane,[34] Langmuir–Blodgett protein thin films,[35] and micro-arrayed multiple cell silicon-devices[33,36] are a few examples.

The presence of a solid substrate may improve nucleation rate through the combination of different aspects: modification of the supersaturation profile nearby the surface due to concentration polarization; template effects due to adsorption of solutes at the surface; or influence of surface characteristics leading to specific interactions with the solute. Most proteins tend to adsorb more extensively at hydrophobic than at hydrophilic surfaces[37,38]; the behavior of proteins undergoing limited or no interfacial conformational changes (so-called "hard" proteins) follows this expectation. This is the case of a reasonably spherical/symmetric proteins such as lysozyme, with approximate dimensions of $3.0 \, nm \times 3.0 \, nm \times 4.5 \, nm$.[39] The extent of solution-membrane interactions depends on pH, since molecules change their charged state as a consequence of modifications in their solution environment. Protein absorption is also affected by the ionic strength. At working $pH < pI$ it is expected that at higher ionic strength protein adsorption is depressed at electrostatically attractive conditions, thus leading to increasing contact angles.

Materials of diverse chemical nature with irregular rough surface structures and pores are designed to act as heterogeneous nucleants in protein crystallization. Some of them include porous silicon,[40–42] bio-glass[41] semi-synthetic micromica,[43] cellulose and hydroxyapatite powders,[42,44] and microporous zeolite.[45] Usually, the wide range of pore sizes and shapes that are available at the surface of disordered porous material provides a large repertoire of pores among which the given macromolecule will be likely to find a pore of adequate size and shape to nucleate.[41] In these materials, the cavities are expected to physically entrap the protein molecules, thus allowing solute concentration up to the required level of supersaturation, not available in the bulk solution, and able to promote nucleation and crystal growth (Fig. 6.10). Accordingly, they might enable crystallization — at metastable conditions — of difficult-to-crystallize proteins, and in some cases without the

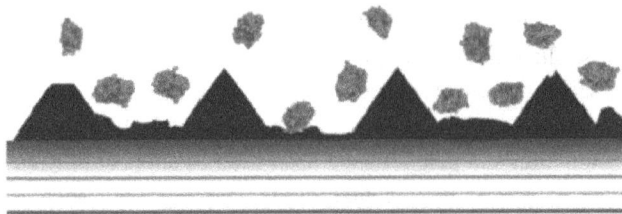

Figure 6.10. Physical entrapment of protein molecules inside the cavities on substrates with irregular (porous) surface topography. This effect allows solute concentrations up to the required level of supersaturation able to promote nucleation.

need for specifically adapted conditions and irrespective of molecular weight, pH, precipitant, and crystallization setup. Interestingly, protein crystallization in the pore permits the formation of different crystal forms, suggesting that pore crystallization may follow a trajectory through the phase diagram that is different from that inherent to the hanging drop method.[42]

The combination of all these aspects is generally observed when using irregular porous membranes as solid templates for heterogeneous nucleation processes in MAC. The chemical and physical nature of the membrane surface influences the mechanisms of solute interaction/aggregation with important effects on the crystallization processes, so that accelerated crystallization kinetics with lower induction times than normal,[22,46–48] but with preserved diffracting crystal quality, are generally observed with MAC.[49–51] In Fig. 6.11 are reported SEM micrographs of typical porous membrane surfaces with tetragonal lysozyme and orthorhombic trypsin crystals embedded on the polymeric surface.

Kinetic data for various PVDF membranes characterized by different morphologies (Table 6.1), measured for HEWL solutions with different precipitant concentrations, are cumulatively plotted in Fig. 6.12. A qualitative increase of the nucleation rate with porosity is observed, in agreement with the tendency of the nucleation barrier ΔG to decrease at higher porosity (see Chapter 2).

Because of its dependence on the contact angle, heterogeneous nucleation rate can be controlled by modulating the extent of interactions between protein solution and membrane. The maximum of the free energy change associated with cluster formation, corresponding to the energy barrier that nuclei must overcome to become stable, is plotted in Fig. 6.13 at different membrane porosity using Eq. (2.42). Under the constraints imposed by the α values, the effect of porosity appears relevant as the energetic barrier decreases at higher porosity. For instance, the predicted $\Delta G_{het}/\Delta G_{hom}$ ratio for PKF-P2 membrane ($\varepsilon = 0.11$) is 0.30, 35% lower than the value calculated by Eq. (2.36) for an ideal dense polymeric matrix

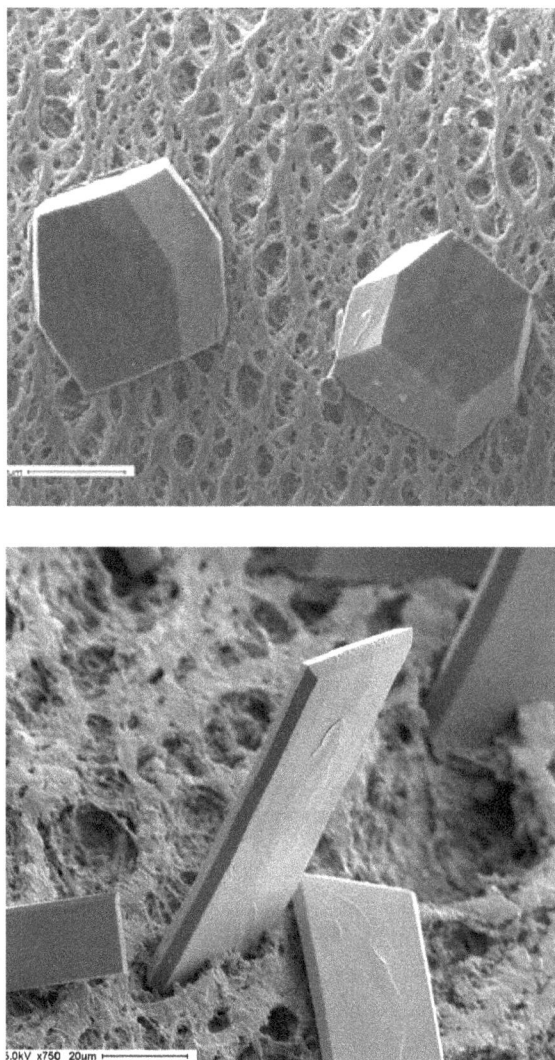

Figure 6.11. SEM images of polypropylene membrane surfaces. Up: embedded tetragonal lysozyme [Reprinted from Di Profio, G., Curcio, E., Cassetta, A. *et al.* (2003). Membrane crystallization of lysozyme: kinetic aspects, *J. Cryst. Growth*, 257, 359–369. Copyright 2014, with permission from Elsevier]; down: trypsin crystals. (Adapted with permission from Di Profio, G., Perrone, G., Curcio, E. *et al.* (20050. Preparation of enzyme crystals with tunable morphology in membrane crystallizers, *Ind. Eng. Chem. Res.*, 44, 10005–10012. Copyright 2014 American Chemical Society).

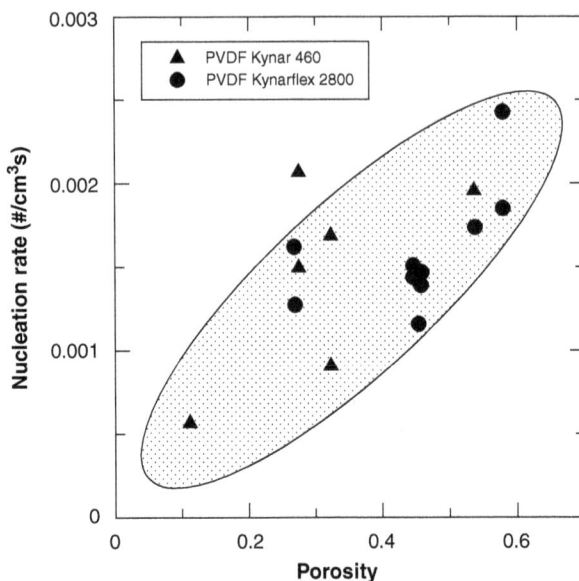

Figure 6.12. The increasing trend of the HEWL heterogeneous nucleation rate with porosity for modified PVDF membranes (precipitant: NaCl; concentration: 2.5% w/v–3% w/v; buffer: AcH/AcNa 0.5 M pH 4.6). (Adapted with permission from Curcio, E., Fontananova, E., Di Profio, G. *et al.* (2006). Influence of the structural properties of poly(vinylidene fluoride) membranes on the heterogeneous nucleation rate of protein crystals, *J. Phys. Chem. B*, 110, 12438–12445. Copyright 2014 American Chemical Society).

having the same contact angle ($87.4° \pm 5.8°$). In the case of the low-porosity PKF membrane ($\varepsilon = 0.0059$), also characterized by the highest measured contact angle ($100° \pm 2.5°$), the ratio $\Delta G_{het}/\Delta G_{hom}$ differs only by 1.5% with respect to the value of 0.64 predicted by the traditional Volmer equation. This confirms the favorable effect of porosity with respect to crystallization kinetics, as the stress energy in the forming crystals can be drastically reduced due to pores existing on the substrate surface.[52]

6.7 Energetics of Protein Nucleation on Polymeric Membranes

The possibility to modulate the physico-chemical properties of the surfaces (such as, hydrophobic character, porosity, roughness, etc.) and to operate at lower supersaturation with respect to those required to activate homogeneous nucleation enhances the probability of obtaining crystals with appropriate size and high structural order from reduced amounts of protein.

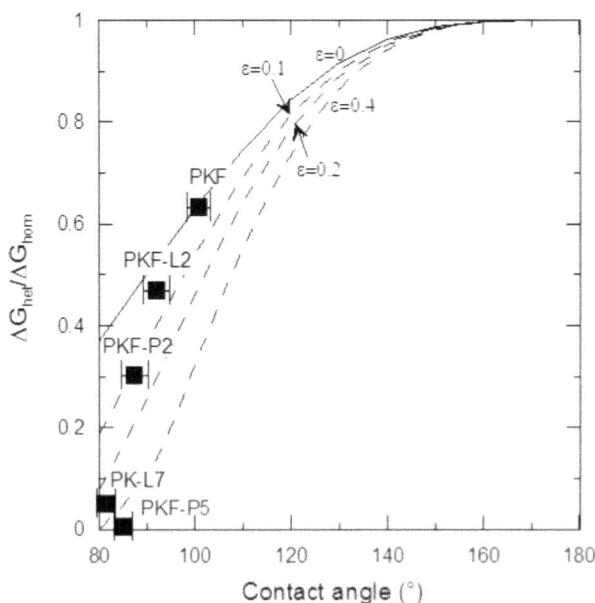

Figure 6.13. $\Delta G_{het}/\Delta G_{hom}$ ratio as a function of the contact angle of HEWL solution (2% w/v NaCl) at different porosity (lines are from Eq. (2.42), symbols are experimental points). (Adapted with permission from Curcio, E., Fontananova, E., Di Profio, G. *et al.* (2006). Influence of the structural properties of poly(vinylidene fluoride) membranes on the heterogeneous nucleation rate of protein crystals, *J. Phys. Chem. B*, 110, 12438–12445. Copyright 2014 American Chemical Society).

As an alternative approach to classical nucleation theory (CNT),[20] Monte Carlo simulation of the Ising model can be used to investigate collective effects occurring in the nucleation of a new phase undergoing first-order phase transition. The 2D Ising model has been successfully applied to study heterogeneous nucleation on square lattice with nearest-neighbor interactions and free boundary conditions.[53–55] The Metropolis Monte Carlo algorithm can be implemented for studying the Ising energy function of a finite square lattice to investigate the energy of heterogeneous nucleation on rough polymeric membranes and studying the interaction between crystallization kinetics and the geometrical profile of the surface.

6.7.1 *Ising model*

We consider a 2D Ising model on a finite square lattice $\Lambda = \{1 \ldots N\}^2$ with spin σ in the configuration $\sigma \in \Omega$, with $\Omega = \{-1, +1\}$. The energy function of the

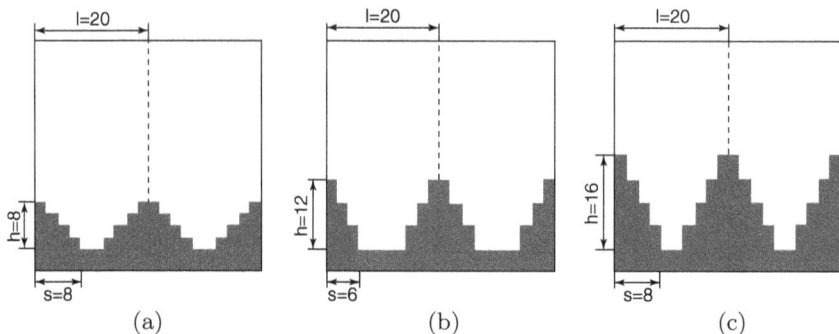

Figure 6.14. Ising lattices of 40 × 40 sites. Black sites, identifying surface asperities, with fixed spins-up $\sigma_I = +1$; the bulk (white sites) is in the spin-down phase $\sigma = -1$. The three initial configurations correspond to different surface roughness coefficients: a) $r = 1.20$; b) $r = 1.33$; c) $r = 1.59$. (Adapted with permission from Curcio, E., Curcio, V., Di Profio, G. et al. (2010). Energetics of protein nucleation on rough polymeric surfaces, J. Phys. Chem. B, 114, 13650–13655. Copyright 2014 American Chemical Society).

system for spin-spin and spin-field interactions is

$$E = -J \sum_{ij} \sigma_i \sigma_j - h \sum_i \sigma_i - J_s \sum_{ij} \sigma_i \sigma_j, \qquad (6.8)$$

where $\sigma_{i,j} = \pm 1$ is the state of spin, J is the coupling constant between free neighboring spins, h the external magnetic field, and J_s is the strength of the coupling between a free spin and a fixed spin of the wall. The surface is formed of a region where spins are fixed to $\sigma_i = +1$ (see Fig. 6.14). It is here assumed that $J_s = 0$, i.e. that the polymeric surface does not preferentially attract either phase; justification arises when considering free energies of surface/spin-up and of surface/spin-down phases identical for a contact angle of 90°, a value not too far from those measured experimentally for polymeric surfaces.

All the simulations described use a lattice of size $L = 40$ and a coupling constant $J = 0.7 kT$, where k is the Boltzmann constant, larger than the critical point of the 2D Ising model at $J = 0.44kT$[56]; a low external magnetic field $h = 0.05 kT$ is also included. The parameter J is set in agreement with previous studies,[54,57] and is consistent with the physical-chemical characteristics of HEWL, chosen as the model system. The exact expression of the interfacial tension γ between the bulk spin-up and spin-down phases in a 2D Ising model with zero external field is given by[56]

$$\frac{\gamma}{kT} = 2\frac{J}{kT} - \ln \left[\frac{\left(1 + \exp\left(-2\frac{J}{kT}\right)\right)}{\left(1 - \exp\left(-2\frac{J}{kT}\right)\right)} \right]. \qquad (6.9)$$

At $J/kT = 0.7$, the predicted value of γ is $0.9kT$ per site. Assuming that the size of each site corresponds to the molecular diameter of HEWL ($\sim 3\,\text{nm}$[58]), the surface tension is 0.5 mN/m, not so far from the value of 1 mN/m that is the HEWL solid-liquid interfacial tension.[59] Simulation makes use of the standard Metropolis Monte Carlo method for spin flipping: flip is always accepted if it lowers the energy or, otherwise, accepted with probability of $\exp(-\Delta E/kT)$, where ΔE is the energy difference due to spin flip, and k is the Boltzmann constant.[60]

The forward flux sampling (FFS) algorithm[61,62] can be used to manage the rare nucleation events. The homogeneous nucleation rate per site and per cycle of $3.2 \pm 0.2 \times 10^{-13}$, is in good agreement with literature results.[54,57] For $l = 40$, the nucleation rate is $3.1 \pm 0.2 \times 10^{-13}$, thus ensuring that the box size is sufficient to make computed events substantially independent of L. However, the number of heterogeneous nucleation rates on rough surfaces is high enough ($>9 \times 10^{-7}$) to be evaluated also by direct simulation. A time sequence of computer simulation snapshots of the 40×40 Ising lattice corresponding to a surface with roughness coefficient of 1.59 is reported in Fig. 6.15.

The rough profile of the surface offers preferential sites for spin flipping from the down to the up phase (Fig. 6.15a), whereas, in the bulk of the system, isolated up spins appear to not be susceptible to aggregation in clusters. This is a clear indication of the enhanced nucleation rate in heterogeneous systems with respect to homogeneous ones. Molecules are trapped in the concavity of rough surfaces, and migration of physically adsorbed molecules is limited between asperities of the membrane surface. This results in a local packing of molecules that enhances the probability of nucleation. After the complete flips of sites at the membrane interface (at 2.0×10^3 cycles·sites), the system remains almost stable in the spin-down phase, before the coalescence of up spins and aggregation in clusters is suddenly enhanced at the valley bottom (Fig. 6.15b). This observation is explained by the existence of an activation barrier for valley filling that occurs at $\sim 1.9 \times 10^4$ cycles·sites (Fig. 6.15c); this value can be interpreted as the induction time τ_{Ising} required by the system to leave the metastable state. In the nucleation regime, only one cluster grows and engulfs the whole lattice, and the induction time is inversely proportional to the nucleation rate N_{Ising} ($N_{Ising} = \tau_{Ising}^{-1}$); therefore, for $r = 1.59$, $N_{Ising} = 5.2 \times 10^{-5}$ #/cycle·site. The filling process is quite symmetrical, coherent with the peculiar geometry of the system. The site flipping is then propagated to the bulk phase; the spin-up number increases steadily to the complete shifting of all the box sites from down to up. Systems with lower roughness coefficients were characterized by earlier appearance of the activation barrier, detected at 4.8×10^4 cycles·sites for $r = 1.33$, and at 1.1×10^6 cycles·sites for $r = 1.20$.

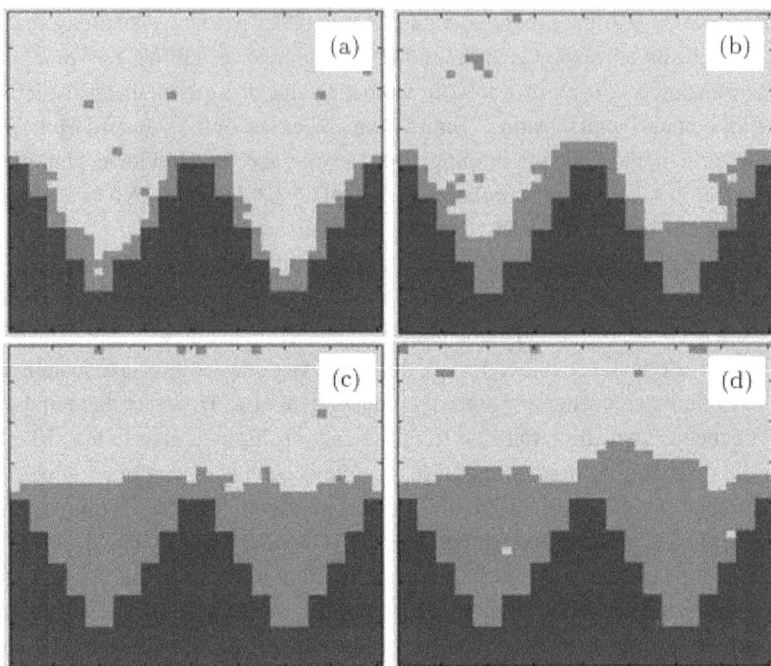

Figure 6.15. A sequence of snapshots of the 40×40 lattice box emulating a surface with roughness coefficient of 1.59 ($J/kT = 0.7, h/kT = 0.05$) at: (a) 2.0×10^3 cycles·sites; (b) 1.8×10^4 cycles·sites; (c) 1.9×10^4 cycles·sites; (d) 2.0×10^4 cycles·sites. The fixed spins are black, up spins are gray, and spins down are light gray. (Adapted with permission from Curcio, E., Curcio, V., Di Profio, G. *et al.* (2010). Energetics of protein nucleation on rough polymeric surfaces, *J. Phys. Chem.B*, 114, 13650–13655. Copyright 2014 American Chemical Society).

Nucleation kinetics have been investigated by evaluating the size of the critical cluster, n^*, and the corresponding energy barrier to nucleation, ΔE^*. Homogeneous nucleation was simulated by setting up a cluster of n spins ($50 \leq n \leq 350$) with $\sigma = +1$ in the middle of the 2D lattice surrounded by a sea of spins with $\sigma = -1$; heterogeneous nucleation on the rough polymeric membrane was simulated by setting up a cluster of n spins ($50 \leq n \leq 350$) with $\sigma = +1$ placed at the valley bottom of the surface profile ($\sigma = +1$) and all remaining spins at $\sigma = -1$. Then, by observing the evolution stories of many individual clusters, we define n^* as the number of up spins of that cluster having $50 \pm 5\%$ probability to growth or decay in 100 runs (each run consisting of 105 steps) of Monte Carlo stochastic dynamics. According to CNT, n^* maximizes the excess internal energy of a nucleus, ΔE, that is defined in the Ising model as the difference in the coupling

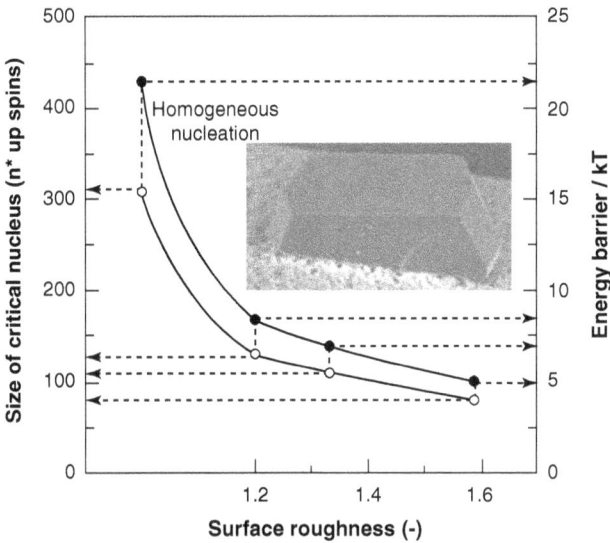

Figure 6.16.   Critical cluster size (n^*) and energy barrier ($\Delta E^*/kT$) of the critical cluster in homogeneous phase and as a function of surface roughness. Inset is a tetragonal HEWL crystal nucleated on the rough surface of a polymeric membrane.

part of the internal energy (the first term in Eq. (6.8)) between a system containing one cluster in a sea of down spins and a homogeneous system (all spins down).[63]

Theoretical results reported in Fig. 6.16 show that, for $J = 0.7kT$ and $h = 0.05kT$, the value of energy barrier for homogeneous nucleation is $\sim 21.5kT$.[58] These findings confirm that nucleation proceeds at low supersaturation, resulting in a single growing large cluster of up spins. Measurements of the distribution of cluster sizes in a 64×64 2D Ising lattice at $J/kT = 0.53$ and $h/kT = 0.08$ resulted in a maximum value of free energy of $\sim 28kT$.[64]

Both the size and the free energy of the critical cluster are reduced in the heterogeneous phase in the presence of a rough surface; in this case, the probability and frequency of nucleation events are enhanced at higher roughness coefficient. In particular, the energy barrier is decreased by 60% (from $21.5kT$ to $8.5kT$) for nuclei growing on a surface with $r = 1.2$; the critical size is consequently reduced from 310 to 130 up spins. A further decrease of both n^* and ΔE^* is observed at increasing r, and the lowest energy barrier ($\sim 5kT$) and cluster size (80 up spins) is found for the most rough simulated profile ($r = 1.59$).

Figure 6.17 compares Ising model simulation results, valid for $\theta = 90°$, and CNT predictions at the same contact angle and at different roughness coefficients. In general, the agreement is satisfactory although the Ising model seems

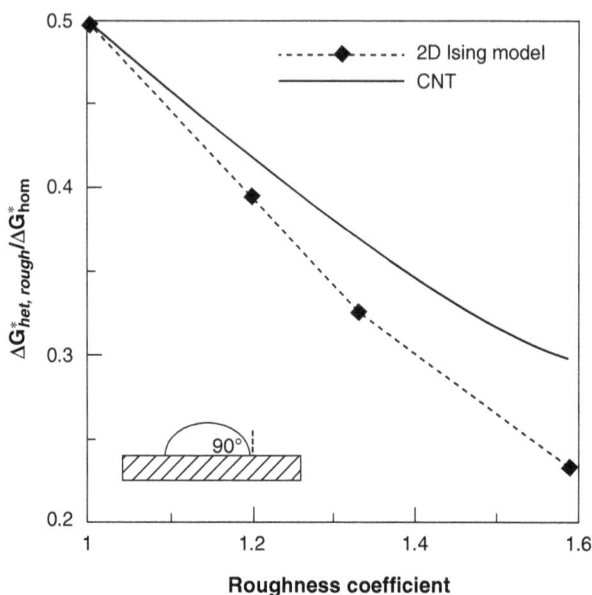

Figure 6.17. Comparison between Ising model and CNT results for $\Delta G^*_{het,rough}/\Delta G^*_{hom}$ as a function of roughness coefficient. Data refers to a contact angle of 90°. (Adapted with permission from Curcio, E., Curcio, V., Di Profio, G. *et al.* (2010). Energetics of protein nucleation on rough polymeric surfaces, *J. Phys. Chem. B*, 114, 13650–13655. Copyright 2014 American Chemical Society).

to underestimate the reduction of energy barrier to heterogeneous nucleation; for moderately rough surfaces ($r < 1.25$), the discrepancy is lower than 10%, whereas deviation is greater than 20% for $r > 1.55$.[65]

A qualitative comparison between theoretical and experimental nucleation rate was also attempted. In fact, protein crystallization is a rather complex process, influenced not only by the physico-chemical and structural characteristics of the surface, but also by electrostatic interactions between protein molecules and polymeric membranes; preferential adsorption of macromolecules at hydrophobic surfaces rather than hydrophilic ones also significantly affects the heterogeneous nucleation kinetics. Moreover, the Ising model permits the evaluation of the nucleation rate only up to an unknown proportionally factor, and it can be expressed in the unit of number of critical nuclei formed per Monte Carlo step and lattice site. As from Fig. 6.18, the values of N_{Ising} span over two orders of magnitude from 9.1×10^{-7} #/cycle·site at low roughness ($r = 1.2$) to 5.2×10^{-5} #/cycle·site at $r = 1.59$.[54] The increasing trend of N_{Ising} with r is experimentally confirmed by crystallization tests on lysozyme. The measured HEWL nucleation rate using PVDF membrane, characterized by a roughness coefficient

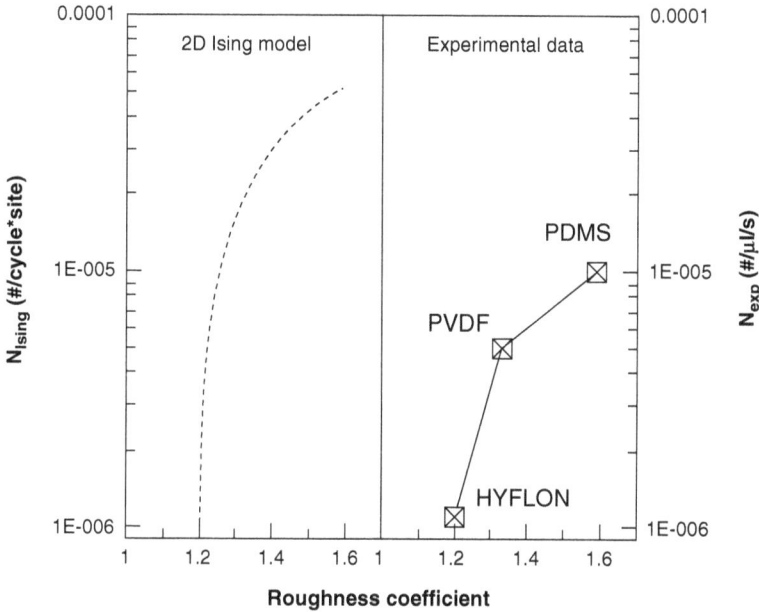

Figure 6.18. Theoretical and experimental nucleation rate as a function of the roughness coefficient. (Adapted with permission from Curcio, E., Curcio, V., Di Profio, G. *et al.* (2010). Energetics of protein nucleation on rough polymeric surfaces, *J. Phys. Chem. B*, 114, 13650–13655. Copyright 2014 American Chemical Society).

of 1.25, is 5×10^{-6} #/μL·s; this is five times higher than the value measured on HYFLON membranes ($r = 1.1$). The faster nucleation process, occurring at a rate of 1×10^{-5} #/μL·s, has been detected on polydimethylsiloxane (PDMS) membranes ($r = 1.5$). Induction times, reduced for higher nucleation rates, were between 1 and 3 hours.[70]

6.8   Growing Proteins in Forced Solution Convection

One of the most controversial aspects regarding protein crystallization is the role of the forced solution convection on the crystallization kinetics. The process of crystal growth occurs according to sequential stages: transport of molecule from the bulk of the mother liquor to the crystal surface; adsorption and diffusion over the surface; attachment and diffusion along a step; and integration into the crystal at a kink site. In solution-based crystal growth, a depletion zone near the solid crystal surface originates from the solidification of material out of the liquid into the solid state;

therefore, a gradient of concentration is produced. This results in density changes that give rise to solute buoyancy-driven convection in the presence of gravity. The slow supply of molecules causes a reduction of protein concentration in the vicinity of the growing faces, thus preventing the formation of new nuclei and giving to the protein molecules a better chance to reach a specific position in the lattice. Moreover, the depletion zone around a growing crystal can help to improve its quality by working as a filtration zone for impurities.[66] This description suggests that solution flow affects the crystallization process in a detrimental way[67–70]; the reduction of the buoyancy-driven convection by using, for example, microgravity, gels, and oils as dispersing media, or by the presence of magnetic/electric fields,[70–74] is considered as a factor improving the quality of protein crystals because, in the absence of fluid convection, molecules are transported toward the crystal by diffusion. However, crystal quality improvement in microgravity was observed only in a few cases,[75] precluding this assumption being taken as a general rule.[76]

On the contrary, macromolecular crystallization in conditions of axial flow, in a laminar regime, is found to improve the crystallization process, thus leading to high-quality crystals.[77–83] Nucleation is enhanced and, at an optimal flow rate (v), maximum nucleation can be attained.[77,80,83] In terms of crystal quality, this can be interpreted by considering the mechanism of solute transport during crystallization in a forced solution regime. In the case of crystallization in forced solution flow conditions, the growth can be governed by a transport and/or by a kinetic control mechanism. In the first case, the crystal growth rate depends on the rate for the solute molecules to be supplied to the growing surface from the bulk of the solution; since convection will increase the rate of solute supply, it is expected that increasing solution velocity will increase crystal growth rate. In the second case, growth rate is governed by the rate of the molecules to be included on the growing face from the interface solid/solution; in these conditions it has been observed that flow does not affect the interfacial kinetics of solute incorporation into the crystals.[84]

By examining the lysozyme growth process under steady solution conditions, the growth mechanism can be described as a combination of both surface kinetics and bulk-transport-based control. In this case, the intrinsic instability of the growth process leads to a fluctuation in growth rate, step density, and step velocity ($\sim80\%$ of their respective average values), thus inducing a decrease in the homogeneity of the crystals, with the appearance of striae on the crystal surface and reduced crystal quality.[85] The strongest instability is expected for processes that operate under equal influence of the transport and kinetics components on the overall control rate, in the same manner as in still conditions.[85] Working under pure kinetic or transport control mechanism, starting from a mixed condition (e.g. by forced solution flow), should result in a more steady growth, thus obtaining higher crystal quality because of the significantly reduced fluctuation amplitudes in growth rate, step density, and step velocity. As a result, the stability of the overall growth process and thus crystal quality is improved.[78,79]

6.8.1 Crystal growth in dynamic membrane crystallizers

When diffusive phenomena in solution are the rate-determining step for crystal growth, a concentration gradient between the bulk and the crystal surface is established; in this case, Fick's first and second laws can be used to describe the diffusion of a molecule, through the boundary layer to a stationary particle. In conditions of forced solution flow, diffusion is enhanced by convective mass transport, and the particle flux J toward a crystal growing in supersaturated environment can be expressed as

$$J = k_x(C_\infty - C_{eq}) = \frac{D}{\delta}(C_\infty - C_{eq}), \tag{6.10}$$

where k_x is the mass transfer coefficient, D the diffusion coefficient, δ the boundary layer thickness, C_∞ and C_{eq} are the bulk and surface (equilibrium) concentrations, respectively. This yields a convective growth rate G given by

$$G = \xi^{-1} k_x c_e (S - 1). \tag{6.11}$$

ξ is the number of molecules precipitated per unit volume of solution and σ the supersaturation ratio ($\sigma = C_\infty/C_{eq}$). The boundary layer thickness δ is evaluated by using Carlson analysis of laminar flow with a slip velocity v[86]:

$$\delta = \left(\frac{x}{0.282}\right) Re^{-1/2} Sc^{-1/3} = \left(\frac{x}{0.282}\right) \left(\frac{vx}{\eta}\right)^{-1/2} \left(\frac{\eta}{D}\right)^{-1/3}. \tag{6.12}$$

x can be approximated by the crystal linear dimension, η is the kinematic viscosity, Re the Reynolds number and Sc the Sherwood number. From this theory, δ is predicted to be inversely proportional to the square root of the velocity and G should increase proportionally to $v^{1/2}$.

The Peclèt number is the control parameter for the individuation of the dominant mechanism and it is defined as the ratio of the rate constants for non-linear integration kinetics and bulk transport in the form of

$$Pe_k = \beta_f \frac{\delta}{D}, \tag{6.13}$$

where β_f is the face kinetic coefficient (e.g. $\sim 10^{-6}$ cm/s for lysozyme). Decreasing values of Pe_k correspond to kinetic controlled regime with significantly faster transport.

6.8.2 Effect of the forced solution convection on the crystal growth rate

In Fig. 6.19 the crystal length and width growth rate (G_l and G_w) as a function of the recirculating solution velocity (v) for bovine trypsin (BPT) crystallized with PEG 6000 (BPTPEG) and porcine trypsin crystallized with ammonium sulfate (PPTAS) are shown; their behaviors have the typical bell shape.[50,77] The same curves obtained for BPT crystallized with ammonium sulfate (BPTAS) are reported in the figure. In the latter two cases, the maximum values for G_l and G_w are reached for the same v of 895 μm/s and 932 μm/s for BPTAS and PPTAS respectively. When BPT is crystallized with PEG 6000, the maximum for G_l is shifted up to $v = 1498\,\mu$m/s, while the G_w maximum is at about $v = 1116\,\mu$m/s. The increase in the average growth rate by increasing v arises from the enhancement in the convective supply of the growing units to the solid-liquid interface.[76] The achievement of a maximum in the curves of Fig. 6.19 is due to the shifting of the operative system working point to a regime with significantly faster transport and higher relative weight of interfacial kinetics effects in the overall control of the crystallization process.[50] Peclèt number calculations confirmed this shifting.[50] Namely, the increase in v induced a decrease in the boundary layer thickness, δ, proportional to $v^{-1/2}$ (Eq. (6.13)), resulting in a decrease of the Peclèt number, with the consequent switch from a diffusive to an interfacial kinetics control mechanism on the overall mass-transfer process.

However, it is evident from Eq. (6.13) that δ is also proportional to the kinematics viscosity ($\delta \rightarrow \eta^{1/6}$) of the solution so that the increase in η would allow the shift of the maximum observed in the G versus v plots towards higher values of v for solutions with an increased kinematics viscosity. The measured values of kinematics viscosity for the three trypsin solutions are: $\eta_{BPTPEG} = 1.92 \times 10^{-2}\,cm^2/s$; $\eta_{PPTAS} = 0.82 \times 10^{-2}\,cm^2/s$; $\eta_{BPTAS} = 0 \times 10^{-2}\,cm^2/s$. These values reflect the order of the maximum of the G_l and G_w curves for the three systems BPTPEG > HEWL > PPTAS $\approx$ BPTAS.

The decrease of G by increasing v after the maximum is explained by the combined effect of impurity[50] and excess of nucleation because of the high solute concentrations in the boundary layer close to the solid–liquid interface, which is actually the case of a kinetic control-based mechanism, which lead to secondary nucleation[87] and suppresses growth. An excess of nucleation is also induced by the increased supersaturation achieved for high solution recirculation velocities. As v increases, the mass-transfer coefficient in the membrane interface rise too (Chapter 2). This generates a higher rate of solvent extraction (high transmembrane flux), which consequently leads to an increased supersaturation that induces an excess of nucleation over the growth. Accordingly, Fig. 6.20 shows big crystals

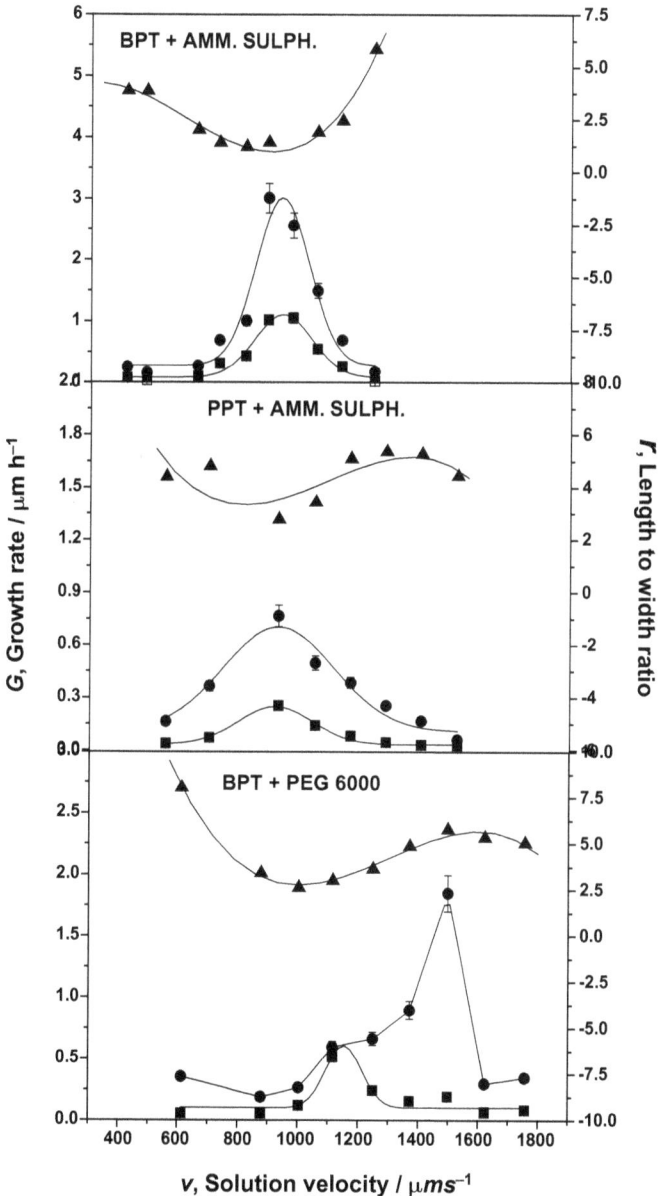

Figure 6.19. Crystal width G_w (■) and length G_l (•) growth rate and length to width ratio r (▲) as a function of the solution velocity (v), for the three protein systems investigated: (a) BPT + ammonium sulfate; (b) PPT + ammonium sulfate; (c) BPT + PEG 6000 (solid lines represent a guide for the reader). (Adapted with permission from Di Profio, G., Perrone, G., Curcio, E. *et al.* (2005). Preparation of enzyme crystals with tunable morphology in membrane crystallizers, *Ind. Eng. Chem. Res.*, 44, 10005–10012. Copyright 2014 American Chemical Society).

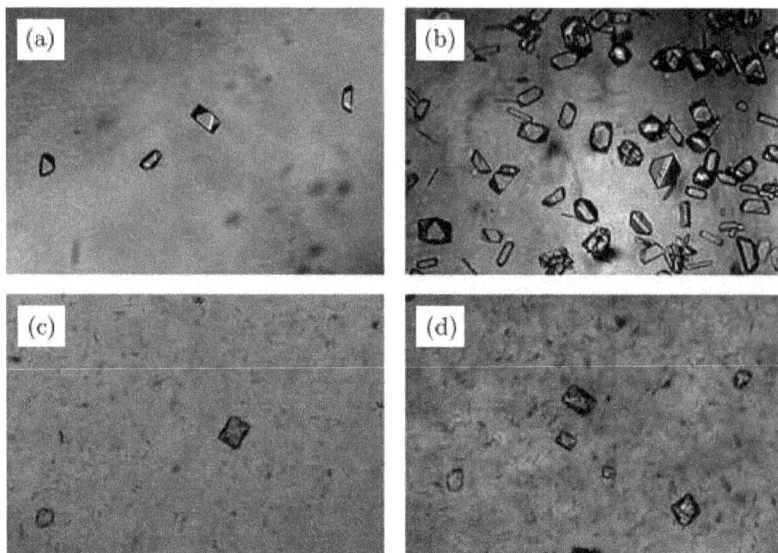

Figure 6.20. BPT crystals obtained in forced solution flow configuration with ammonium sulfate as precipitant, at: (a) 423 μm/s; (b) 821 μm/s; (c) 1072 μm/s; (d) 1139 μm/s. (Adapted with permission from Di Profio, G., Perrone, G., Curcio, E. *et al.* (2005). Preparation of enzyme crystals with tunable morphology in membrane crystallizers, *Ind. Eng. Chem. Res.*, 44, 10005–10012. Copyright 2014 American Chemical Society).

surrounded by a huge amount of smaller tiny crystals that, in the hypothesis, are generated by an excess of nucleation for increasing recirculating velocities.

References

1. McPherson, A. (2004). Introduction to protein crystallization, *Methods*, **34**, 254–265.
2. McPherson, A. (1991). A brief history of protein crystal growth, *J. Cryst. Growth*, **110**, 1–10.
3. McPherson, A. (1999). *Crystallization of Biological Macromolecules*, Cold Spring Harbor Laboratory Press, Cold Spring Harbor.
4. Vellard, M. (2003). The enzyme as drug: Application of enzymes as pharmaceuticals, *Curr. Opin. Biotech.*, **14**, 444–450.
5. Margolin, A.L. and Navia, M.A. (2001). Protein crystals as novel catalytic materials, *Angew. Chem. Int. Ed.*, **40**, 2204–2222.
6. St. Clair, N.L. and Navia, M.A. (1992). Cross-linked enzyme crystals as robust biocatalysts, *J. Am. Chem. Soc.*, **114**, 7314–7316.
7. Falkner, J.C., Al-Somali, A.M., Jamison, J.A. *et al.* (2005). Generation of size-controlled, submicrometer protein crystals, *Chem. Mater.*, **17**, 2679–2686.
8. Margolin, A.L., Khalaf, N.K., St. Clair, N.L. *et al.* (2003). Stabilized protein crystals, formulations containing them and methods of making them, US175239.

9. Vilenchik, L.Z., Griffith, J.P., St. Clair, N. *et al.* (1998). Protein crystals as novel microporous materials, *J. Am. Chem. Soc.*, **120**, 4290–4294.
10. Cvetkovic, A., Picioreanu, C., Straathof, A.J.J. *et al.* (2005). Relation between pore sizes of protein crystals and anisotropic solute diffusivities, *J. Am. Chem. Soc.*, **127**, 875–879.
11. Doye, J.P.K., Louis, A.A. and Vendruscolo, M. (2004). Inhibition of protein crystallization by evolutionary negative design, *Phys. Biol.*, **1**, 9–13.
12. Datta, R., Dechapanichkul, S., Kim, J.S. *et al.* (1992). A Generalized Model for the Transport of Gases in Porous, Non-porous, and Leaky Membranes. I. Application to Single Gases, *J. Membr. Sci.*, **75**, 245–263.
13. Grant, M.L. (2000). Effects of thermodynamic nonideality in protein crystal growth, *J. Crystal Growth*, **209**, 130–137.
14. Haas, C., Drenth, J., Wilson, W.W. *et al.* (1999). Relation between the solubility of proteins in aqueous solutions and the second virial coefficient of the solution, *J. Phys. Chem. B*, **103**, 2808–2811.
15. Guo, B., Kao, S., McDonald, H. *et al.* (1999). Correlation of second virial coefficients and solubilities useful in protein crystal growth, *J. Cryst. Growth*, **196**, 424–433.
16. George, A. and Wilson, W.W. (1994). Predicting protein crystallization from a dilute solution property, *Acta Crystallogr. D*, **50**, 361–365.
17. Gunton, J.D., Shiryayev, A. and Pagan, D.L. (2007). *Protein Condensation: Kinetic Pathways to Crystallization and Disease*, Cambridge University Press, Cambridge, England.
18. Bhamidi, V., Varanasi, S. and Schall, C.A. (2002). Measurement and modeling of protein crystal nucleation kinetics, *Cryst. Growth Des.*, **2**, 395–400.
19. Curcio, E., Di Profio, G. and Drioli, E. (2002). A new membrane-based crystallization technique: Tests on lysozyme, *J. Crystal Growth*, **247**, 166–176.
20. Volmer, M. and Weber, A. (1926). Keimbildung in übersättigten Gebilden, *Z. Phys. Chem.*, **119**, 227–301.
21. Dao, L.H., Nguyen, H.M. and May, H.H. (2000). A fiber optic turbidity system for in-situ monitoring protein aggregation, nucleation and crystallization, *ACTA Astronautica*, **47**, 399–409.
22. Hu, H., Hale, T., Yang, X. *et al.* (2001). A spectrophotometer-based method for crystallization induction time period measurement, *J. Cryst. Growth*, **232**, 86–92.
23. Ansari, R., Suh, K., Arabshashi, A. *et al.* (1996). Fiber Optic Probe for Monitoring Protein Aggregation, Nucleation, and Crystallization, *J. Cryst. Growth*, **168**, 216–226.
24. Chýlek, P. and Li, J. (1995). Light scattering by small particles in an intermediate region, *Opt. Comm.*, **117**, 389–394.
25. Saikumar, M.V., Glatz, C.E. and Larson, M.A. (1998). Lysozyme crystal growth and nucleation kinetics, *J. Cryst. Growth*, **187**, 277–288.
26. Garcìa-Ruiz, J.M. and Moreno, A. (1997). Growth kinetics of protein single crystals in the gel acupuncture technique, *J. Cryst. Growth*, **178**, 393–401.
27. Finet, S., Bonnetè, F., Frouin, J. *et al.* (1998). Lysozyme crystal growth, as observed by small angle X-ray scattering, proceeds without crystallization intermediates, *Eur. Biophys. J.*, **27**, 263–271.
28. Saridakis, E. and Chayen, N.E. (2009). Towards a 'universal' nucleant for protein crystallization, *Trends Biotechnol.*, **27**, 99–106.

29. McPherson, A. and Shlichta, P. (1988). Heterogeneous and epitaxial nucleation of protein crystals on mineral surfaces, *Science*, **230**, 385–387.
30. Kimble, W.L., Paxton, T.E., Rousseau, R.W. *et al.* (1998). The effect of mineral substrates on the crystallization of lysozyme, *J. Cryst. Growth*, **187**, 268–276.
31. Fermani, S., Falini, G., Minnucci, M. *et al.* (2001). Protein crystallization on polymeric film surfaces, *J. Cryst. Growth*, **224**, 327–334.
32. Rong, L., Komatsu, H. and Yoda, S. (2002). Control of heterogeneous nucleation of lysozyme crystals by using poly-L-lysine modified substrate, *J. Cryst. Growth*, **235**, 489–493.
33. Sanjoh, A., Tsukihara, T. and Gorti, S. (2001). Surface-potential controlled Si-microarray devices for heterogeneous protein crystallization screening, *J. Cryst. Growth*, **232**, 618–628.
34. Nanev, C.N. (2007). On the slow kinetics of protein crystallization, *Cryst. Growth Des.*, **7**, 1533–1540.
35. Pechkova, E. and Nicolini, C.J. (2004). Atomic structure of a Ck2alpha human kinase by Microfocus diffraction of extra small microcrystals grown with nanobiofilm template, *J. Cell. Biochem.*, **91**, 1010–1020.
36. Sanjoh, A. and Tsukihara, T. (1999). Spatiotemporal protein crystal growth studies using microfluidic silicon devices, *J. Cryst. Growth*, **196**, 691–702.
37. Norde, W. (1986). Adsorption of proteins from solution at the solid-liquid interface, *Adv. Colloid Interface Sci.*, **25**, 267–340.
38. Haynes, C.A. and Norde, W. (1994). Globular proteins at solid/liquid interfaces, *Colloids and Surfaces B*, **2**, 517–566.
39. Malmsten, M. J. (1998). Formation of Adsorbed Protein Layers, *Colloid Interface Sci.*, **207**, 186–199.
40. Chayen, N.E., Saridakis, E., El-Bahar, R. *et al.* (2001). Porous silicon: An effective nucleation-inducing material for protein crystallization, *J. Mol. Biol.*, **312**, 591–595.
41. Chayen, N.E., Saridakis, E. and Sear, R.P. (2006). Experiment and theory for heterogeneous nucleation of protein crystals in a porous medium, *Proc. Nat. Acad. Sci. USA*, **103**, 597–601.
42. Dekker, C., Haire, L. and Dodson, G. (2004). Vapor-diffusion protein crystallization in pore strips, *J. Appl. Crystallogr.*, **37**, 862–866.
43. Takehara, M., Ino, K., Takakusagi, Y. *et al.* (2008). Use of layer silicate for protein crystallization: Effects of micromica and chlorite powders in hanging drops, *Anal. Biochem.*, **373**, 322–329.
44. Thakur, A.S., Robin, G., Guncar, G. *et al.* (2007). Improved success of sparse matrix protein crystallization screening with heterogeneous nucleating agents, *PLoS One*, **2**, 1091.
45. Sugahara, M., Asada, Y., Morikawa, Y. *et al.* (2008). Nucleant-mediated protein crystallization with the application of microporous synthetic zeolites, *Acta Crystallog. D Biol. Crystallogr.*, **64**, 686–695.
46. Drenth, J., Dijkstra, K., Haas, C. *et al.* (2003). Effect of molecular anisotropy on the nucleation of lysozyme *J. Phys. Chem. B*, **107**, 4203–4207.
47. Paxton, T.E., Sambanis, A. and Rousseau, R.W. (2001). Influence of vessel surfaces on the nucleation of protein crystals, *Langmuir*, **17**, 3076–3079.

48. Kulkarni, A.M. and Zukoski, C.F. (2002). Nanoparticle crystal nucleation: Influence of solution conditions, *Langmuir*, **18**, 3090–3099.
49. Di Profio, G., Curcio, E., Cassetta, A. *et al.* (2003). Membrane crystallization of lysozyme: kinetic aspects, *J. Cryst. Growth*, **257**, 359–369.
50. Curcio, E., Simone, S., Di Profio, G. *et al.* (2005). Membrane crystallisation of lysozyme under forced flow, *J. Membr. Sci.*, **257**, 134–143.
51. Curcio, E., Di Profio, G. and Drioli, E. (2003). A new membrane-based crystallization technique: tests on lysozyme, *J. Cryst. Growth*, **247**, 166–176.
52. Luryi, S. and Suhir, E. (1986). A new approach to the high-quality epitaxial growth of lattice-mismatched materials, *Appl. Phys. Lett.*, **49**, 140–142.
53. Cirillo, E.N.M. and Lebowitz, J.L. (1998). Metastability in the two-dimensional Ising model with free boundary conditions, *J. Stat. Phys.*, **90**, 211–226.
54. Sear, R.P. (2006). Heterogeneous and homogeneous nucleation compared: Rapid nucleation on microscopic impurities, *J. Phys. Chem. B*, **110**, 4985–4989.
55. Page, A.J. and Sear, R.P. (2006). Heterogeneous nucleation in and out of pores, *Phys. Rev. Let.*, **97**, 065701.
56. Onsager, L. (1944). A two-dimensional model with an order-disorder transition, *Phys. Rev.*, **65**, 117–149.
57. Allen, R.J., Valeriani, C., Tanase-Nicola, S. *et al.* (2008). Homogeneous nucleation under shear in a two dimensional Ising model: Cluster growth, coalescence, and breakup, *J. Chem. Phys.*, **129**, 134704.
58. Kisler, J.M., Stevens, G.W. and O'Connor, A. (2001). Adsorption of Proteins on Mesoporous. Molecular Sieves, *J. Mater. Phys. Mech.*, **4**, 89–93.
59. Krishnan, C.A., Maheshwari, R. and Dhathathreyan, A. (2006). Solid/liquid interfacial tension as a tool to study stability of lysozyme on adsorption to solid surfaces, *Chem. Phys. Lett.*, **417**, 128–131.
60. Chandler, D. (1987). *Introduction to Modern Statistical Mechanics*, Oxford University Press, New York.
61. Allen, R.J., Frenkel, D. and ten Wolde, P.R. (2006). Forward flux sampling-type schemes for simulating rare events: Efficiency analysis, *J. Chem. Phys.*, **124**, 194111.
62. Allen, R.J., Frenkel, D. and ten Wolde, P.R. (2006). Simulating rare events in equilibrium or nonequilibrium stochastic systems, *J. Chem. Phys.*, **124**, 024102.
63. Vehkamaki, H. and Ford, I.J. (1999). Nucleation theorems applied to the Ising model, *Phys. Rev. E*, **59**, 6483–6488.
64. Brendel, K., Barkema, G.T. and van Beijeren, H. (2005). Nucleation times in the 2D Ising model, *Phys. Rev. E*, **71**, 031601.
65. Curcio, E., Curcio, V., Di Profio, G. *et al.* (2010). Energetics of protein nucleation on rough polymeric surfaces, *J. Phys. Chem. B*, **114**, 13650–13655.
66. Tanaka, H., Inaka, K., Sugiyama, S. *et al.* (2004). Numerical analysis of the depletion zone formation around a growing protein crystal, *Ann. N.Y. Acad. Sci.*, **1027**, 10–19.
67. Pusey, M., Witherow, W. and Naumann, R. (1988). Preliminary investigations into solutal flow about growing tetragonal lysozyme crystals, *J. Cryst. Growth*, **90**, 105–111.
68. Pusey, M.L. (1992). Continuing adventures in lysozyme crystal growth, *J. Cryst. Growth*, **122**, 1–7.
69. Nyce, T.A. and Rosenberger, F. (1991). Growth of protein crystals suspended in a closed loop thermosyphon, *J. Cryst. Growth*, **110**, 52–59.

70. Miller, T.Y., He, X. and Carter, D.C. (1992). A comparison between protein crystals grown with vapor diffusion methods in microgravity and protein crystals using a gel liquid-liquid diffusion ground-based method, *J. Cryst. Growth*, **122**, 306–309.
71. Vergara, A., Lorber, B., Sauter, C. *et al.* (2005). Lessons from crystals grown in the Advanced Protein Crystallisation Facility for conventional crystallisation applied to structural biology, *Biophys. Chem.*, **118**, 102–112.
72. Moreno, A., Quiroz-Garcia, B., Yokaichiya, F. *et al.* (2007). Protein crystal growth in gels and stationary magnetic fields, *Cryst. Res. Technol.*, **42**, 231–236.
73. Ramachandran, N. and Leslie, F.W. (2005). Using magnetic fields to control convection during protein crystallization — analysis and validation studies, *J. Cryst. Growth*, **274**, 297–306.
74. Kadowaki, A., Yoshizaki, I., Adachi, S. *et al.* (2006). Effects of Forced Solution Flow on Protein-Crystal Quality and Growth Process, *Cryst. Growth Des.*, **6**, 2398–2403.
75. DeLucas, L.J., Moore, K.M., Long, M.M. *et al.* (2002). Protein crystal growth in space, past and future, *J. Cryst. Growth*, **237/239**, 1646–1650.
76. Beth, M., Broom, M.B.H., Witherow, W.K. *et al.* (1988). Preliminary observations of the effect of solutal convection on crystal morphology, *J. Cryst. Growth*, **90**, 130–135.
77. Vekilov, P.G. and Rosenberger, F. (1988). Protein crystal growth under forced solution flow: Experimental setup and general response of lysozyme, *J. Cryst. Growth*, **186**, 251–261.
78. Adachi, H., Takano, K., Matsumura, H. *et al.* (2004). Protein crystal growth with a two-liquid system and stirring solution, *J. Synch. Radiat.*, **11**, 121–124.
79. Adachi, A., Matsumura, H., Takano, K. *et al.* (2003). New practical technique for protein crystallization with floating and stirring methods, *Jpn. J. Appl. Phys.*, **42**, L1161–L1163.
80. Kadowaki, A., Yoshizaki, I., Rong, L. *et al.* (2004). Improvement of protein crystal quality by forced flow solution, *J. Synch. Radiat.*, **11**, 38–40.
81. Azadani, A.N., Lopez, J.M. and Hirsa, A.H. (2007). Protein Crystallization at the Air/Water Interface Induced by Shearing Bulk Flow, *Langmuir*, **23**, 5227–5230.
82. Adachi, H., Takano, K., Nino, A. *et al.* (2005). olution stirring initiates nucleation and improves the quality of adenosine deaminase crystals, *Acta Cryst. D*, **61**, 759–762.
83. Penkova, A., Pan, W., Hodjaoglu, F. *et al.* (2006). Nucleation of protein crystals under the influence of solution shear flow, *Ann. N.Y. Acad. Sci.*, **1077**, 214–231.
84. Lin, H., Rosenberger, F., Alexander, J.I.D. *et al.* (1995). Convective-diffusive transport in protein crystal growth, *J. Cryst. Growth*, **151**, 153–162.
85. Vekilov, P.G., Alexander, J.I.D. and Rosenberger, F. (1996). Nonlinear response of layer growth dynamics in the mixed kinetics-bulk transport regime, *Phys. Rev. E*, **54**, 6650–6660.
86. Dirksen, J.A. and Ring, T.A. (1991). Fundamentals of crystallization: kinetic effects on particle size distributions and morphology, *Chem. Eng. Sci.*, **46**, 2389–2427.
87. McPherson, A. (1993). Effects of a microgravity environment on the crystallization of biological macromolecules, *Microgravity Sci. Tech.*, **6**, 101–109.

Index

www.ingramcontent.com/pod-product-compliance
Lightning Source LLC
Chambersburg PA
CBHW050556190326
41458CB00007B/2065